A Guide to Integrated Control of Turfgrass Diseases

Volume I. Cool Season Turfgrasses

by

L.L. Burpee, Ph.D.
University of Georgia

04426358/

Library of Congress Card Catalog Number: 93-73312
ISBN: 0-9635707-1-4

Copyright ©1993 by L.L. Burpee

All rights reserved.
No part of this book may be reproduced in any form, including photocopy, microfilm, information storage and retrieval system, computer database or software, or by any means, including electronic or mechanical, without written permission from the author.

Published by the
GCSAA Press
Golf Course Superintendents Association of America
1421 Research Park Drive
Lawrence, KS 66049-3859
(785) 841-2240

TURFGRASS
INFO CTR.

SB
608
.T87
B8
1993
v.1
c.4

Use of trade names in this book is for the purpose of illustration only and should not be construed as a product endorsement or condemnation. Lists of fungicides are not meant to be comprehensive. All chemicals used for pest control should be applied in strict compliance with local, state, and federal regulations. Prior to the use of any pesticide, the applicator must comply with all current regulations.

THIS BOOK FOR REFERENCE ONLY

READ THE LABEL ON ALL PESTICIDES CAREFULLY

Dedicated to the memory of

Clifford G. Warren
Turfgrass Pathologist

PREFACE

This book contains the information necessary to develop integrated strategies for the management of turfgrass diseases. The information is presented in the form of disease triangles and disease management triangles for each of the major species of cool season turfgrasses. All pertinent information is presented on two pages for each disease. Each disease triangle is on a left-hand (even numbered) page and each disease management triangle is on a right-hand (odd numbered) page. Turfgrass managers are encouraged to use information from all sides of a disease management triangle to develop an integrated approach to disease control.

The contents of this book are arranged under headings that coincide with the major species of cool season turfgrasses. This arrangement was chosen because most turfgrass managers are concerned with controlling diseases on only one or two species of turfgrasses. They often do not have the time to read extensive information about disease control in general. Therefore, turf managers who are familiar with the species of grasses that they maintain can use this book to locate specific information on disease management quickly and efficiently.

This book emphasizes control of turfgrass diseases rather than diagnosis. An excellent source of diagnostic information is the Compendium of Turfgrass Diseases, 2nd Edition, by R.W. Smiley, P.H. Dernoeden and B.B. Clark, APS Press, St. Paul, MN (ISBN: 0-89054-124-8).

L.L. Burpee
Department of Plant Pathology
University of Georgia
Georgia Station
Griffin, GA 30223

CONTENTS

PAGE

Fine-Leaf Fescues

FOREWORD

More than seventy diseases are known to afflict turfgrasses. Most of these diseases are caused by living organisms that are capable of inciting disease under rather specific and narrow ranges of environmental conditions. Plant pathologists have used this information to define disease as an interaction among a causal agent (pathogen), a plant (host) and the specific environmental conditions required for the pathogen to infect and incite disease in the host. This three-way interaction can be conceptualized in the form of a triangle - a disease triangle (Figure 1).

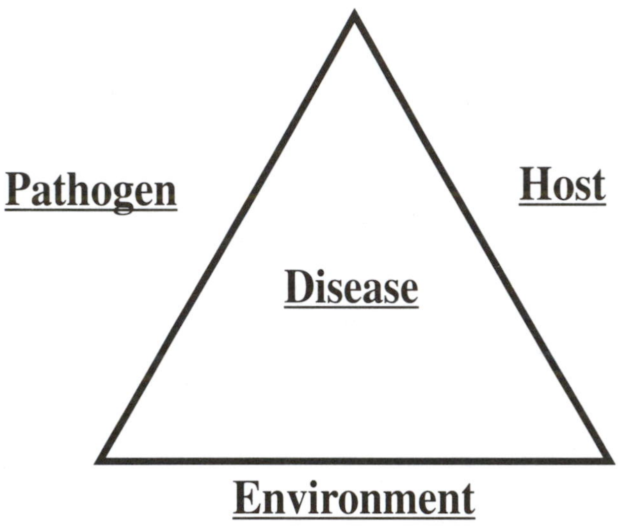

Figure 1. A disease triangle.

As a corollary to the disease triangle, disease management or control can be conceptualized as a triangle where manipulations of pathogens, hosts and environmental conditions lead to the prevention or alleviation of disease (Figure 2). Pathogens are manipulated by applying fungicides and nematicides to turfgrasses or soils. Manipulation of the host is achieved by replacing a susceptible turfgrass species or cultivar with a disease resistant turfgrass. The turfgrass environment can be manipulated by incorporating changes in management practices such as the amount and timing of irrigation, the frequency of topdressing or the height of mowing. These are examples of cultural controls.

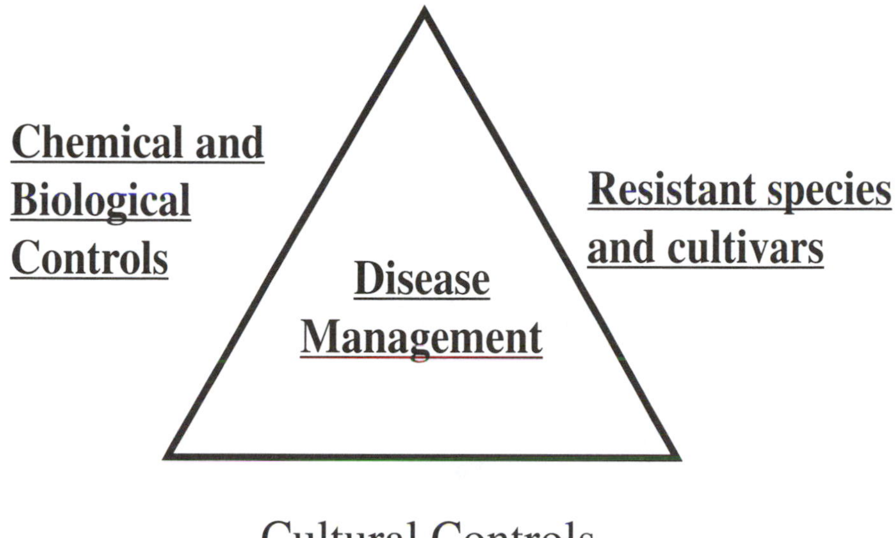

Figure 2. A disease management triangle.

This book includes disease triangles and disease management triangles for the major diseases of cool season turfgrasses. In addition, sources of disease forecasting and pathogen monitoring units are printed above each disease management triangle. Disease forecasting materials include equipment for monitoring environmental parameters such as temperature and relative humidity, which are analyzed by a computer program to estimate when a disease will occur. Pathogen monitoring equipment includes devices designed to detect and identify pathogenic fungi in turfgrass plants. These devices can assist in diagnosis and in timing control practices such as fungicide applications.

Turfgrass diseases are relatively short-lived. They may be active for only a few days or weeks but result in destruction of large areas of turf. Battling a disease while it is actively spreading is often futile. By the time that symptoms are visible, the pathogen has frequently established latent infections that are often impossible to stop. Knowledgeable turfgrass managers know that the best strategies for disease management are those that are focused on prevention. They also know that the use of multiple strategies for disease control may prevent problems such as fungicide resistance and the development of new races of pathogens. The disease triangles and disease management triangles in this book will provide a foundation for establishing long-term integrated control of turfgrass diseases.

New information about the biology and management of turfgrass diseases is constantly being disseminated. Undoubtedly, by the time that this book comes to press there will be new facts relevant to particular diseases. Therefore, each triangle has been designed with extra space, in the form of blank lines, that the reader can use to add pertinent new information about a particular disease.

Chapter 1

Bentgrasses and Annual Bluegrass

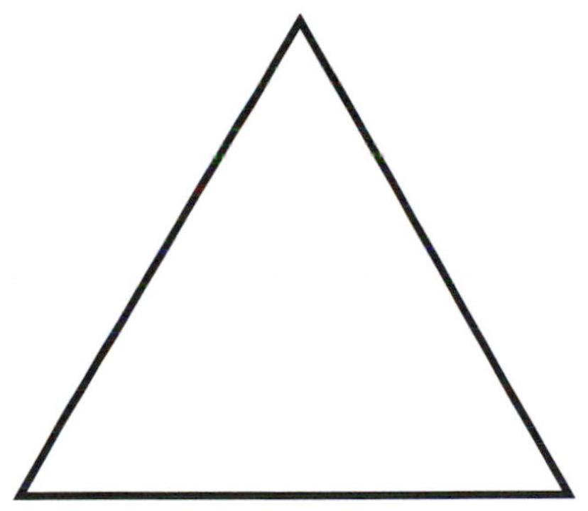

Anthracnose

Hosts

Creeping bentgrass, bluegrasses, fescues, perennial ryegrass. Annual bluegrass is particularly susceptible.

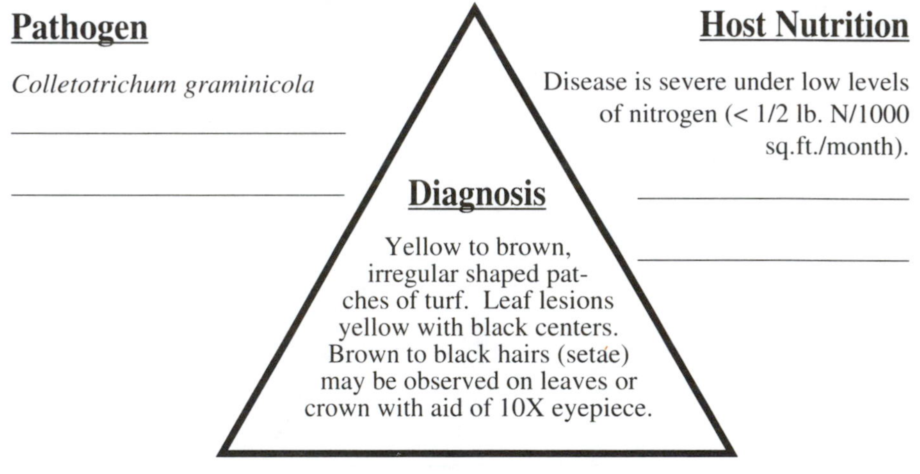

Pathogen

Colletotrichum graminicola

Host Nutrition

Disease is severe under low levels of nitrogen (< 1/2 lb. N/1000 sq.ft./month).

Diagnosis

Yellow to brown, irregular shaped patches of turf. Leaf lesions yellow with black centers. Brown to black hairs (setae) may be observed on leaves or crown with aid of 10X eyepiece.

Environment

Temperature >78°F (26°C).
More than 10 hrs. of leaf wetness per day for several days.
Disease is particularly severe on turf exposed to soil compaction and excess thatch.
Pathogen may cause crown rot of creeping bentgrass at temperatures from 60°-77° F (16°-26°C).

Anthracnose

Disease Forecasting and Pathogen Detection

An anthracnose forecaster is available from Neogen Corp., 620 Lesher Place, Lansing, MI 48912. Detection kits are not available.

Chemical Controls

Banner, Bayleton, Cleary's 3336, ConSyst, Daconil, Dithane, Fungo-Flo, Lesco Systemic and Granular, Mancozeb, Rubigan*, Scott's (Fluid Fungicide, Fungicide III, VII, Systemic Fungicide), Twosome.

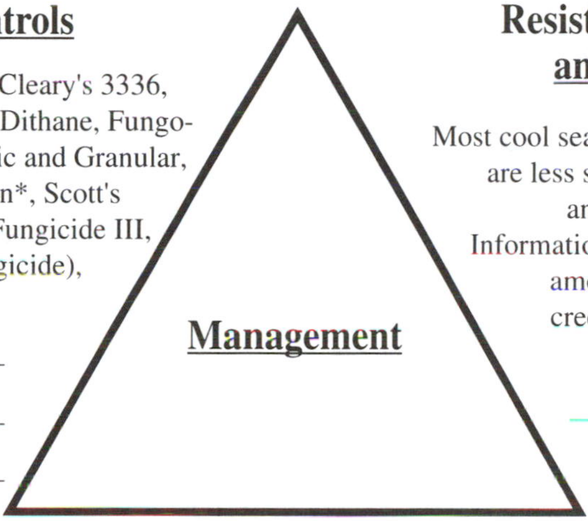

Resistant Species and Cultivars

Most cool season turfgrasses are less susceptible than annual bluegrass. Information on resistance among cultivars of creeping bentgrass is limited.

Management

Cultural Controls

Applications of 1/2 lb. N/1000 sq.ft./month reduce disease severity.
Use light-weight mowing equipment (reduce compaction).
Limit thatch thickness to 1/4 inch or less.
Decrease shade and increase air circulation to enhance drying of turf.
Syringe turf with water when temperature >80°F (27°C).
Avoid irrigation in late afternoon and in evening prior to midnight.

* May reduce populations of annual bluegrass.

Bacterial Wilt

Hosts

Specific strains of the pathogen infect creeping bentgrass, other strains infect annual bluegrass. Toronto C-15, Seaside and Nimisilia are particularly susceptible cultivars of bentgrass.

Pathogen

Xanthomonas campestris

Host Nutrition

Effects of turf nitrition on disease are unknown.

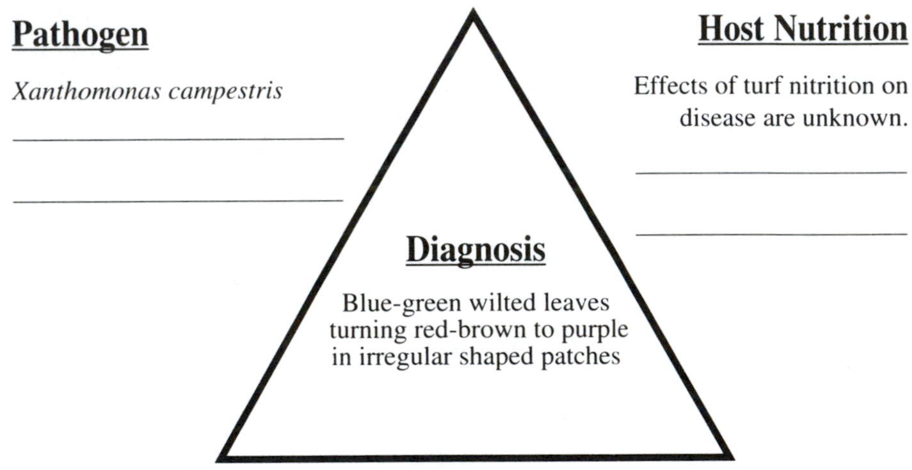

Diagnosis

Blue-green wilted leaves turning red-brown to purple in irregular shaped patches

Environment

Disease is severe during warm to cool, wet periods and at low mowing heights (< 1/2 inch).

Bacterial Wilt

Disease Forecasting and Pathogen Detection

A bacterial wilt forecaster and detection kits are not available.

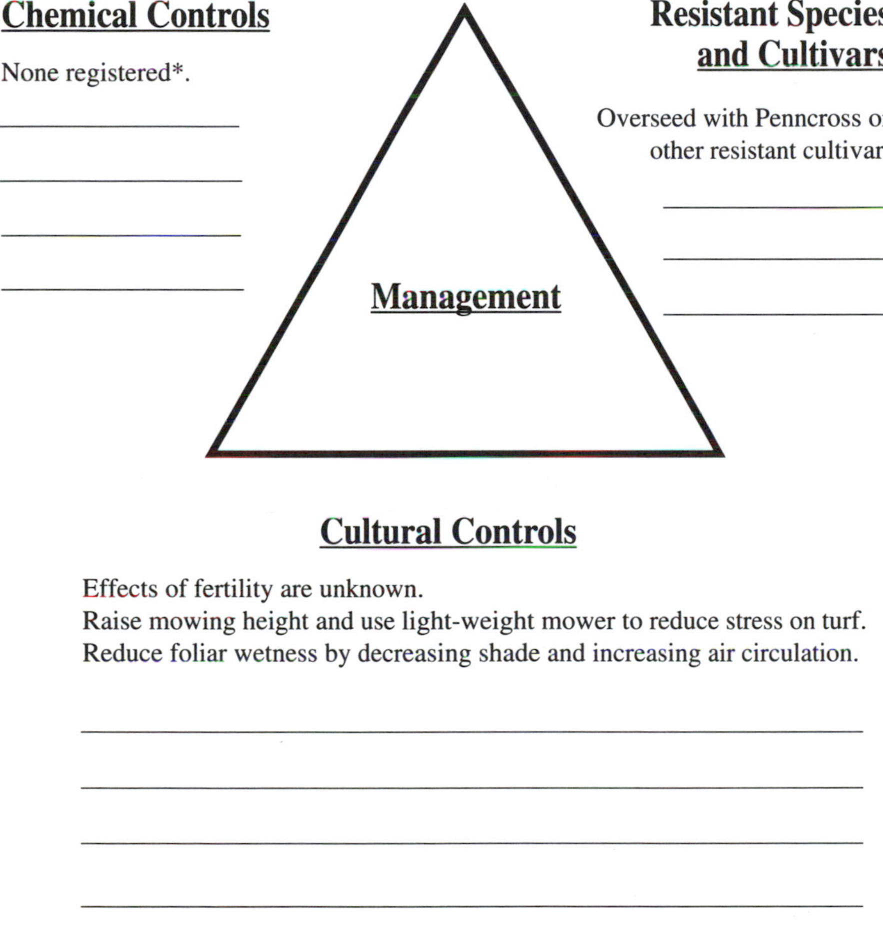

Chemical Controls

None registered*.

Resistant Species and Cultivars

Overseed with Penncross or other resistant cultivar.

Cultural Controls

Effects of fertility are unknown.
Raise mowing height and use light-weight mower to reduce stress on turf.
Reduce foliar wetness by decreasing shade and increasing air circulation.

* Mycoshield (C. Pfizer Corp.) has been used experimentally to control this disease.

Brown Patch

(Rhizoctonia Blight)

Hosts

All common species of turfgrasses.

Pathogen

Rhizoctonia solani

Diagnosis

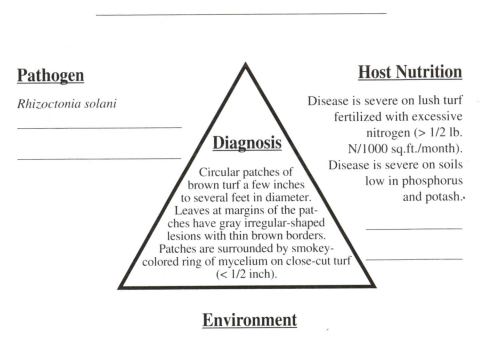

Circular patches of brown turf a few inches to several feet in diameter. Leaves at margins of the patches have gray irregular-shaped lesions with thin brown borders. Patches are surrounded by smokey-colored ring of mycelium on close-cut turf (< 1/2 inch).

Host Nutrition

Disease is severe on lush turf fertilized with excessive nitrogen (> 1/2 lb. N/1000 sq.ft./month). Disease is severe on soils low in phosphorus and potash.

Environment

Night temperatures >60°F (16°C).
More than 10 hrs. of foliar wetness per day for several days.
Disease is severe at low mowing heights (< 1/2 inch).

Brown Patch

Disease Forecasting and Pathogen Detection

A brown patch forecaster and detection kits are available from Neogen Corp., 620 Lesher Pl., Lansing, MI 48912.

Chemical Controls

Banner, Bayleton, Benomyl, Captan, Chipco 26019, Cleary's 3336, ConSyst, Curalan, Daconil, Dithane, Duosan, Fungo, Fore, Lesco Systemic and Granular, Mancozeb, PCNB*, Prostar, Rubigan**, Scott's (Fluid Fungicide, Fungicides II, III, VII, IX, X, Systemic Fungicide, FFII*), Spotrete, Terraclor*, Thiram, Touché, Turfcide*, Twosome.

Management

Resistant Species and Cultivars

Creeping bentgrass is less susceptible than colonial or velvet bentgrass. The bentgrass cultivars Pro/Cup, Providence, Cobra and National are less susceptible than other cultivars.

Cultural Controls

Maintain moderate nitrogen fertility (1/2 lb. N/1000 sq.ft./month).
Maintain moderate phosphorous and high potash according to soil tests.
Decrease shade and increase air circulation to enhance drying of turf.
Avoid irrigation in late afternoon and in evening prior to midnight.
Maintain thatch at 1/4 inch thick or less.
Raise mowing height if possible.

* Avoid application to bentgrasses. ** May reduce populations of annual bluegrass.

Copper Spot

Hosts

Creeping bentgrass and other species of bentgrass.
Velvet bentgrass is particularly susceptible.

Pathogen

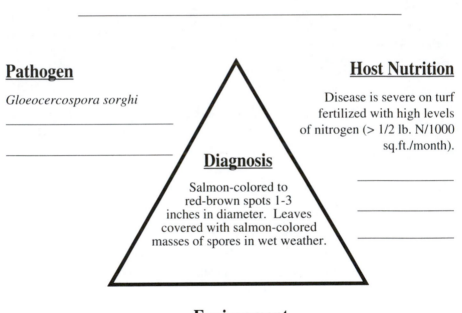

Gloeocercospora sorghi

Host Nutrition

Disease is severe on turf fertilized with high levels of nitrogen (> 1/2 lb. N/1000 sq.ft./month).

Diagnosis

Salmon-colored to red-brown spots 1-3 inches in diameter. Leaves covered with salmon-colored masses of spores in wet weather.

Environment

Temperature 78°- 86°F (26°- 30°C).
More than 10 hrs. of leaf wetness per day for several days.
Soil pH < 6.

Copper Spot

Disease Forecasting and Pathogen Detection

A copper spot forecaster and detection kits are not available.

Chemical Controls

Bayleton, Cleary's 3336, ConSyst, Daconil, Dithane, Duosan, Fore, Fungo, Lesco Systemic and Granular, Mancozeb, Rubigan*, Scott's (Fungicide IX, Systemic Fungicide), Twosome.

Management

Resistant Species and Cultivars

Creeping bentgrass is less susceptible than velvet bentgrass. Information on resistant cultivars is limited.

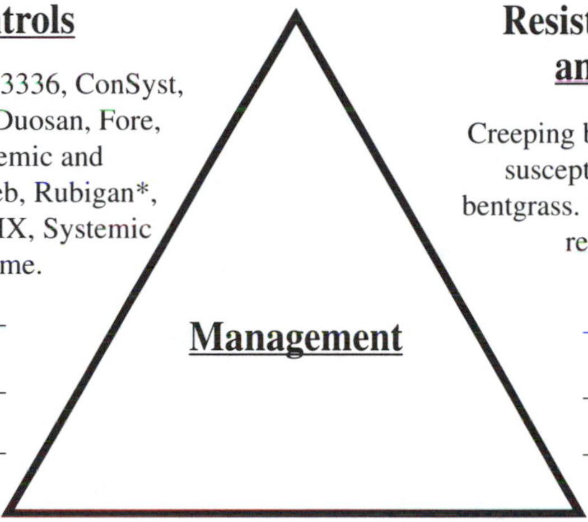

Cultural Controls

Avoid applying > 1/2 lb. N/1000 sq.ft./month during spring and summer.
Decrease shade and increase air circulation to enhance drying of turf.
Apply lime to achieve a pH of 6-7.

* May reduce populations of annual bluegrass.

Curvularia Blight

Hosts

Bentgrasses, bluegrasses and perennial rye.

Pathogen

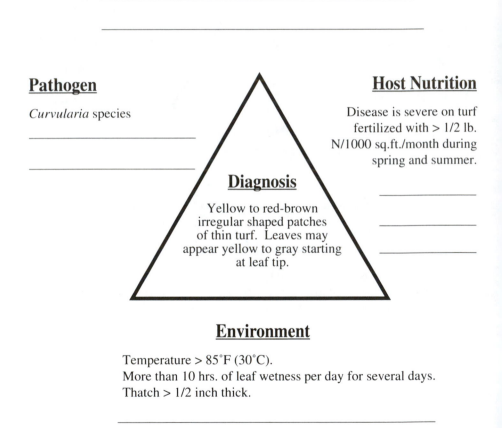

Curvularia species

Host Nutrition

Disease is severe on turf fertilized with > 1/2 lb. N/1000 sq.ft./month during spring and summer.

Diagnosis

Yellow to red-brown irregular shaped patches of thin turf. Leaves may appear yellow to gray starting at leaf tip.

Environment

Temperature > 85°F (30°C).
More than 10 hrs. of leaf wetness per day for several days.
Thatch > 1/2 inch thick.

Curvularia Blight

Disease Forecasting and Pathogen Detection

A curvularia blight forecaster and detection kits are not available.

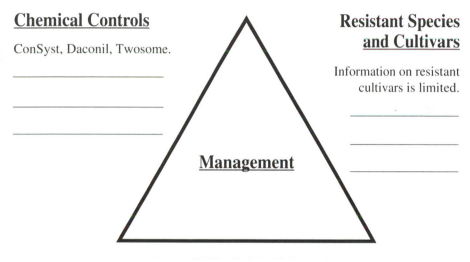

Chemical Controls

ConSyst, Daconil, Twosome.

Resistant Species and Cultivars

Information on resistant cultivars is limited.

Management

Cultural Controls

Fertilize with not more than 1/2 lb. N/1000 sq.ft./month during spring and summer.
Use light-weight mowing equipment (reduce compaction).
Limit thatch thickness to 1/4 inch or less.
Decrease shade and increase air circulation to enhance drying of turf.
Syringe turf with water when temperature > 80˚F (27˚C).
Avoid irrigation in late afternoon and in evening prior to midnight.
Raise mowing height if possible.

Damping-off and Seed Rot

Hosts

All species of turfgrasses.

Pathogen

Species of *Pythium, Fusarium* or *Rhizoctonia*

Host Nutrition

Disease is severe on seed or seedlings fertilized excessively with nitrogen or on turf subjected to nutrient deficiency.

Diagnosis

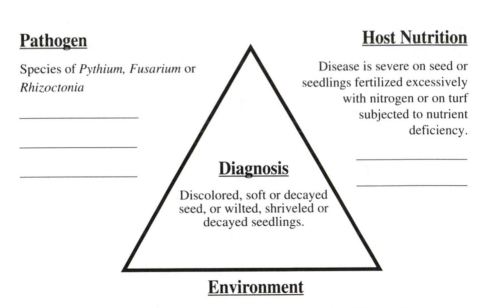

Discolored, soft or decayed seed, or wilted, shriveled or decayed seedlings.

Environment

Temperatures too high (> 85°F, 30°C) or too low (< 60°F, 15° C) for optimum seedling development.
More than 10 hrs. of seed or seedling wetness per day for several days.
Excessive shade or overcrowding of seedlings.
Poor surface or subsurface drainage.

Damping-off and Seed Rot

Disease Forecasting and Pathogen Detection

A damping-off or seed rot forecaster is not available.
Detection kits for ***Pythium*** and ***Rhizoctonia*** are available from Neogen Corp., 620 Lesher Pl., Lansing, MI 48912.

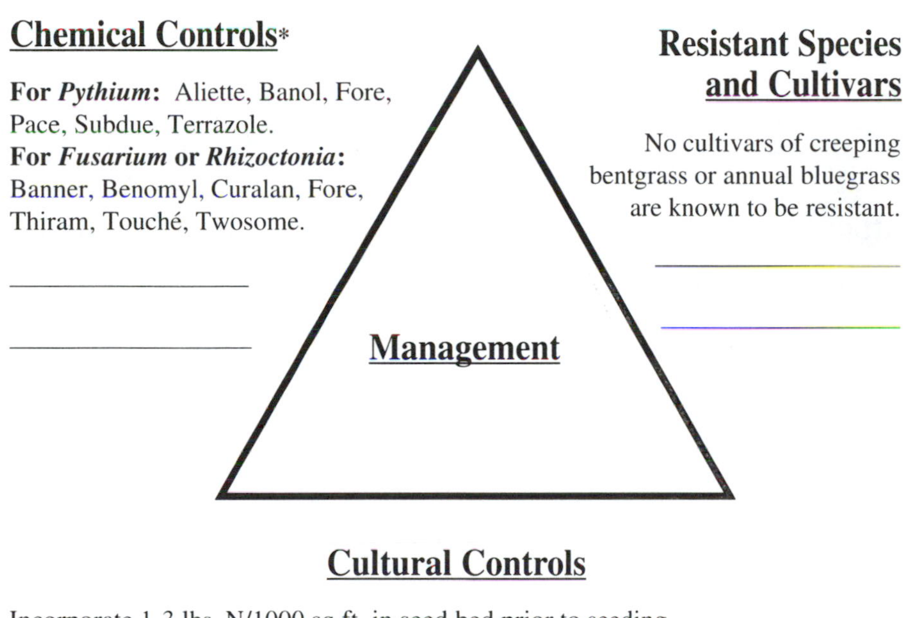

Chemical Controls*

For *Pythium*: Aliette, Banol, Fore, Pace, Subdue, Terrazole.
For *Fusarium* or *Rhizoctonia*: Banner, Benomyl, Curalan, Fore, Thiram, Touché, Twosome.

Resistant Species and Cultivars

No cultivars of creeping bentgrass or annual bluegrass are known to be resistant.

Management

Cultural Controls

Incorporate 1-3 lbs. N/1000 sq.ft. in seed bed prior to seeding.
Incorporate phosphorous and potash according to soil tests.
Fertilize seedlings with 1/2 to 1 lb. N/1000 sq.ft./month.
Seed turf when day temperatures are between 60°- 80°F (15°-27°C).
Use recommended seeding rate (1/2 to 1 lb. seed/1000 sq. ft.) for creeping bentgrass.
Avoid high seeding rates.
Avoid irrigation in late afternoon and in evening prior to midnight.
Reduce shade and increase air circulation to enhance drying of turf.
Raise mowing height and use light-weight equipment. Improve surface and subsurface drainage.

* Some materials may be sold as seed-treatments as well as foliar sprays.

Dollar Spot

Hosts

All common species of turfgrasses.

Pathogen

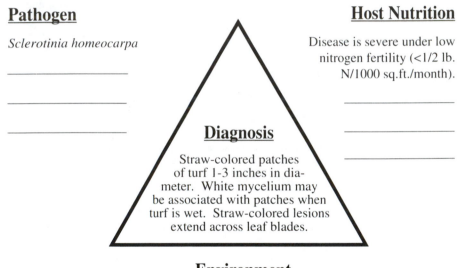

Sclerotinia homeocarpa

Host Nutrition

Disease is severe under low nitrogen fertility (<1/2 lb. N/1000 sq.ft./month).

Diagnosis

Straw-colored patches of turf 1-3 inches in diameter. White mycelium may be associated with patches when turf is wet. Straw-colored lesions extend across leaf blades.

Environment

Night temperatures > 50°F (10°C) and day temperatures < 90°F (32°C).
More than 10 hrs. of leaf wetness per day for several days.
Disease is severe on turf subjected to drought stress.

Dollar Spot

Disease Forecasting and Pathogen Detection

A dollar spot forecaster is available from Pest Management Supply, P.O. Box 938, Amherst, MA 01004. Detection kits are available from Neogen Corp., 620 Lesher Pl., Lansing, MI 48912.

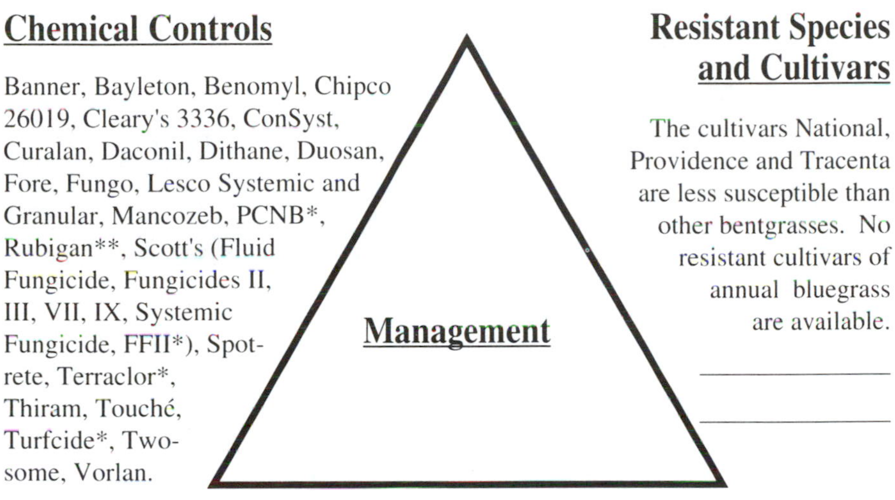

Chemical Controls

Banner, Bayleton, Benomyl, Chipco 26019, Cleary's 3336, ConSyst, Curalan, Daconil, Dithane, Duosan, Fore, Fungo, Lesco Systemic and Granular, Mancozeb, PCNB*, Rubigan**, Scott's (Fluid Fungicide, Fungicides II, III, VII, IX, Systemic Fungicide, FFII*), Spot-rete, Terraclor*, Thiram, Touché, Turfcide*, Two-some, Vorlan.

Resistant Species and Cultivars

The cultivars National, Providence and Tracenta are less susceptible than other bentgrasses. No resistant cultivars of annual bluegrass are available.

Cultural Controls

Applications of 1/2 to 1 lb. of N/1000 sq.ft. every 2-4 weeks will reduce severity of dollar spot.
Maintain moderate to high levels of soil potassium as determined by soil tests.
Limit thatch to 1/4 inch or less.
Decrease shade and increase air circulation to enhance drying of turf.
Avoid irrigation in late afternoon and in evening prior to midnight.
Avoid drought stress.

* Avoid application to bentgrasses. ** May reduce populations of annual bluegrass.

Drechslera Leaf Blight and Crown Rot

Hosts

Creeping bentgrass plus other cool season grasses. Not reported on annual bluegrass. The cultivar Toronto C-15 of creeping bentgrass is highly susceptible.

Pathogen

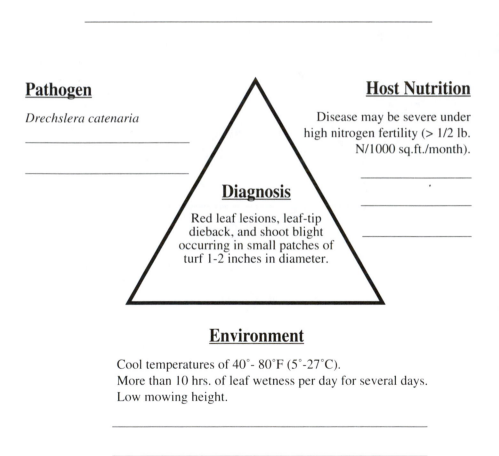

Drechslera catenaria

Host Nutrition

Disease may be severe under high nitrogen fertility (> 1/2 lb. N/1000 sq.ft./month).

Diagnosis

Red leaf lesions, leaf-tip dieback, and shoot blight occurring in small patches of turf 1-2 inches in diameter.

Environment

Cool temperatures of 40°- 80°F (5°-27°C).
More than 10 hrs. of leaf wetness per day for several days.
Low mowing height.

Drechslera Leaf Blight and Crown Rot

Disease Forecasting and Pathogen Detection

A leaf blight and crown rot forecaster and detection kits are not available.

Chemical Controls

No fungicides are registered for this disease, but chemicals registered for Leaf Spot and Melting-out are probably effective.

Resistant Species and Cultivars

Bentgrasses other than Toronto C-15 have some resistance.

Management

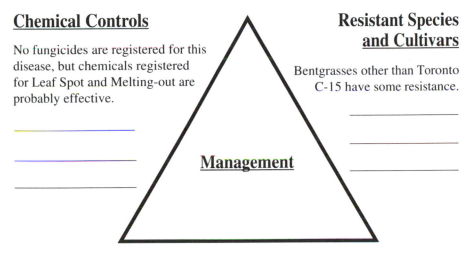

Cultural Controls

Maintain moderate nitrogen fertility (1/2 lb. N/1000 sq.ft./month).
Decrease shade and increase air circulation to enhance drying of turf.
Avoid irrigation in late afternoon and in evening prior to midnight.
Raise mowing height.
Use light-weight mowing equipment to avoid stress on turf.
Limit thatch thickness to 1/4 inch or less.

Fairy Ring

Hosts

All turfgrasses.

Pathogen

Several species of "mushroom-forming" fungi.

Host Nutrition

High nitrogen fertility (> 1/2 lb. N/1000 sq.ft./month) may increase disease severity. Low nitrogen may increase the frequency of occurrence of fairy ring.

Diagnosis

Circles or archs of mushrooms or wilted, dead or dark green turf. White mats of fungal mycelium may be found in thatch or soil associated with circles or archs.

Environment

Light to moderate textured soils.
Soil pH of 5-7.5.
Low to moderate soil moisture.

Fairy Ring

Disease Forecasting and Pathogen Detection

A fairy ring forecaster and detection kits are not available.

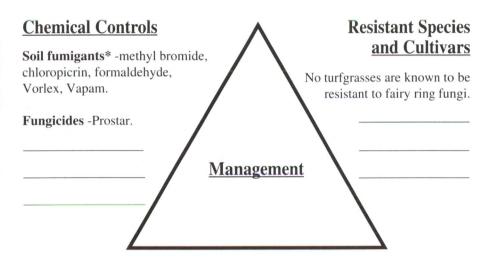

Chemical Controls

Soil fumigants* -methyl bromide, chloropicrin, formaldehyde, Vorlex, Vapam.

Fungicides -Prostar.

____

____

____

Resistant Species and Cultivars

No turfgrasses are known to be resistant to fairy ring fungi.

____

____

____

Management

Cultural Controls

Maintain moderate nitrogen fertility (1/2 lb. N/1000 sq.ft./month).
Maintain moderate to high levels of phosphorus and potash according to soil tests.
Excavate ring and soil 12 inches deep and 24 inches beyond ring or arch and replace with new soil.
Remove sod, cultivate soil 6 to 8 inches deep in several directions, add wetting agent to soil, reseed or sod.

____

____

____

____

* These chemicals are highly toxic to turfgrasses, animals and other life.

Fusarium Patch

Hosts

All cool season turfgrasses. Bentgrasses, annual bluegrass and perennial ryegrass are particularly susceptible.

Pathogen

Microdochium nivale

Host Nutrition

Disease is severe under high nitrogen fertility (> 1/2 lb. N/1000 sq.ft./month) during late summer, fall, winter and spring. Disease may be severe on soils low in phosphorus and potash.

Diagnosis

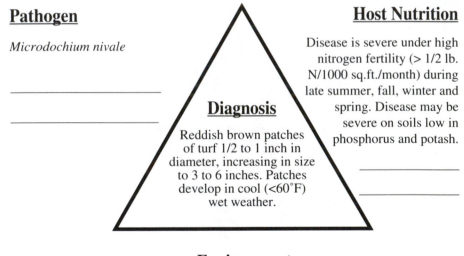

Reddish brown patches of turf 1/2 to 1 inch in diameter, increasing in size to 3 to 6 inches. Patches develop in cool (<60˚F) wet weather.

Environment

Temperature <60˚F (16˚C).
More than 10 hrs. of leaf wetness per day for several days.
Disease severity may be increased by applications of lime.

Fusarium Patch

Disease Forecasting and Pathogen Detection

A Fusarium patch forecaster and detection kits are not available.

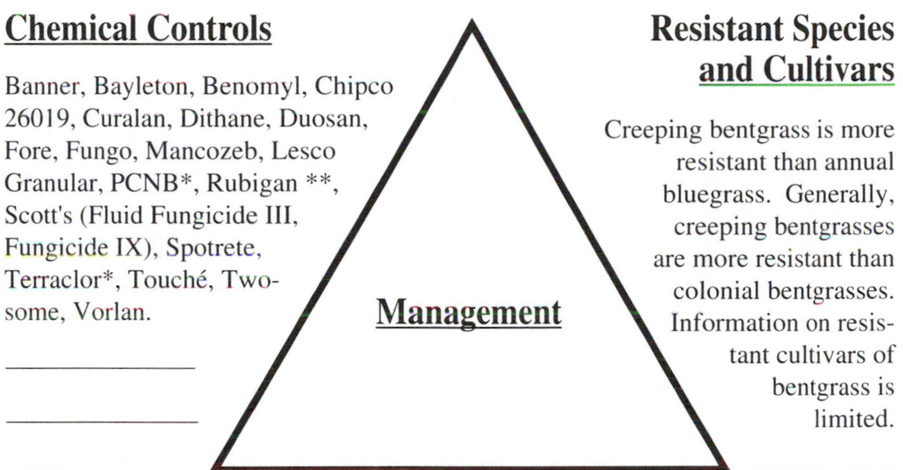

Chemical Controls

Banner, Bayleton, Benomyl, Chipco 26019, Curalan, Dithane, Duosan, Fore, Fungo, Mancozeb, Lesco Granular, PCNB*, Rubigan **, Scott's (Fluid Fungicide III, Fungicide IX), Spotrete, Terraclor*, Touché, Two-some, Vorlan.

Resistant Species and Cultivars

Creeping bentgrass is more resistant than annual bluegrass. Generally, creeping bentgrasses are more resistant than colonial bentgrasses. Information on resistant cultivars of bentgrass is limited.

Cultural Controls

Avoid high nitrogen (> 1/2 lb. N/1000 sq.ft./month) in late summer and early fall.
Maintain moderate to high levels of phosphorus and potash according to soil tests.
Decrease shade and increase air circulation to enhance drying of turf.
Avoid applications of lime if possible.
Avoid irrigation in late afternoon and in evening prior to midnight.

* Avoid application to actively growing bentgrass. ** May reduce populations of annual bluegrass.

Leaf Spot

Hosts

Bluegrasses, bentgrasses, fescues and perennial ryegrass.

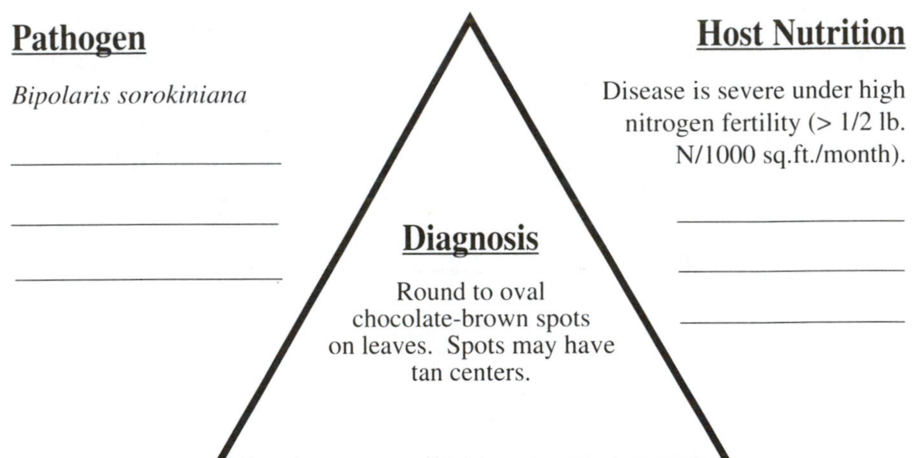

Pathogen

Bipolaris sorokiniana

Host Nutrition

Disease is severe under high nitrogen fertility (> 1/2 lb. N/1000 sq.ft./month).

Diagnosis

Round to oval chocolate-brown spots on leaves. Spots may have tan centers.

Environment

Temperature of 77°- 95°F (25°- 35°C).
Disease severity increases with increases in temperature.
More than 10 hrs. of leaf wetness per day for several days.

Leaf Spot

Disease Forecasting and Pathogen Detection

A leaf spot forecaster and detection kits are not available.

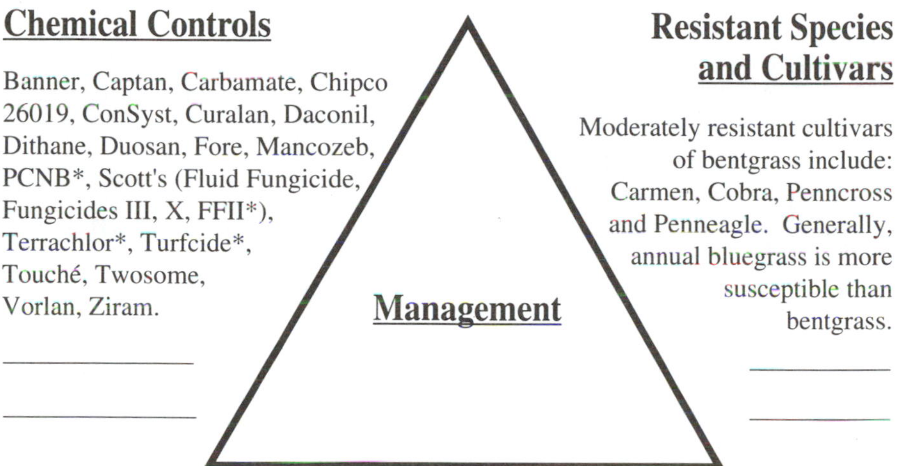

Chemical Controls

Banner, Captan, Carbamate, Chipco 26019, ConSyst, Curalan, Daconil, Dithane, Duosan, Fore, Mancozeb, PCNB*, Scott's (Fluid Fungicide, Fungicides III, X, FFII*), Terrachlor*, Turfcide*, Touché, Twosome, Vorlan, Ziram.

Management

Resistant Species and Cultivars

Moderately resistant cultivars of bentgrass include: Carmen, Cobra, Penncross and Penneagle. Generally, annual bluegrass is more susceptible than bentgrass.

Cultural Controls

Apply moderate amounts of nitrogen during summer (1/4-1/2 lb. N/ 1000 sq.ft./month).
Maintain moderate to high levels of soil P and K.
Decrease shade and increase air circulation to enhance drying of turf.
Avoid irrigation in late afternoon and in evening prior to midnight.
Limit thatch to 1/4 inch or less.
Raise mowing height.
Use light-weight mowing equipment to reduce stress.

* Avoid application to bentgrasses.

Melting-Out

Hosts

Bluegrasses, ryegrasses and tall fescue. This pathogen has not been reported on creeping bentgrass.

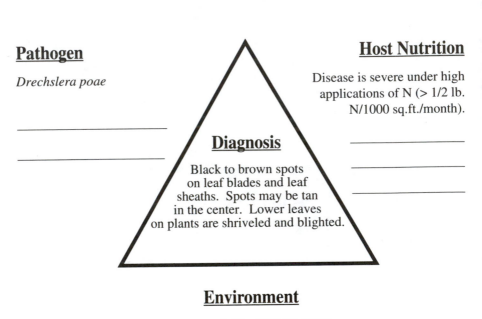

Pathogen

Drechslera poae

Host Nutrition

Disease is severe under high applications of N (> 1/2 lb. N/1000 sq.ft./month).

Diagnosis

Black to brown spots on leaf blades and leaf sheaths. Spots may be tan in the center. Lower leaves on plants are shriveled and blighted.

Environment

Cool temperatures of 40°- 80°F (5°-27°C).
More than 10 hrs. of leaf wetness per day for several days.
Mowing height < 2 inches.

Melting-Out

Disease Forecasting and Pathogen Detection

A melting-out forecaster and detection kits are not available.

Chemical Controls

Banner, Captan, Chipco 26019, ConSyst, Curalan, Daconil, Dithane, Fore, Mancozeb, Scott's (Fluid Fungicide, Fungicide III), Terraclor*, Turfcide*, Touché, Twosome, Vorlan.

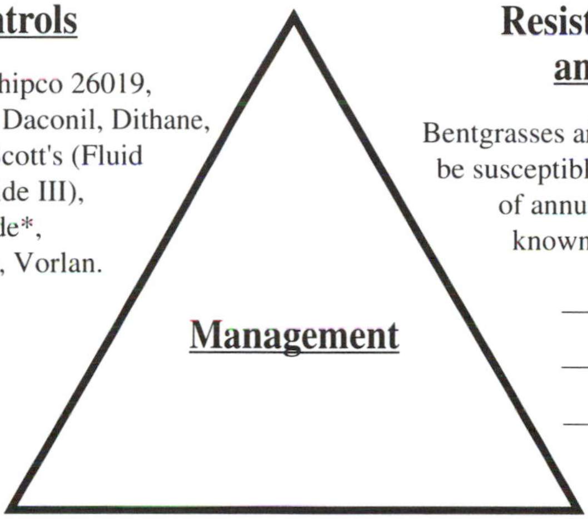

Resistant Species and Cultivars

Bentgrasses are not known to be susceptible. No cultivars of annual bluegrass are known to be resistant.

Management

Cultural Controls

Fertilize with low to moderate levels of N during spring, summer and fall (1/4-1/2 lb. N/1000 sq.ft./month).
Decrease shade and increase air circulation to enhance drying of turf.
Avoid irrigation in late afternoon and in evening prior to midnight.
Raise mowing height.
Use light-weight mowing equipment to avoid stress on turf.
Limit thatch thickness to 1/4 inch or less.

* Avoid application to bentgrasses.

Nematodes

Hosts

All common species of turfgrasses.

Pathogen

More than 25 different species attack cool season turfgrasses.

Host Nutrition

Symptoms are often severe on lush, excessively fertilized turf or on thin, under-fertilized turf.

Diagnosis

Irregular shaped light green or yellow patches of turf up to several feet in diameter. Leaves may be yellow or brown from the tip. Roots may be thin, stunted or knotted.

Environment

Soil temperatures > 40°F (5°C).
Symptoms are often severe on turf growing in sandy, light-textured soils.
Symptoms may be enhanced by drought and high temperatures (> 80°F, 27°C).

Nematodes

Disease Forecasting and Pathogen Detection

A nematode forecaster and detection kits are not available.

Chemical Controls

Post-plant nematicides:
Clandosan 618, Dasanit, Mocap*, Nemacur, Scott's Nematicide/ Insecticide.
Pre-plant nematicides:
Basamid, Brom-o-Sol, Telone, Terr-o-Cide, Vorlex.

Resistant Species and Cultivars

Information on resistance among species of cool season turfgrasses is limited.

Management

Cultural Controls

Maintain a balanced fertility program.
Apply 1/2 lb. N/1000 sq.ft./month during spring and summer.
Maintain moderate to high levels of phosphorus and potash according to soil tests.
Have soil analyzed for nematodes prior to seeding or sodding.
Use sod that is nematode-free.

* Avoid applications to bentgrasses.

Pink Patch

Hosts

Perennial ryegrass, fine-leaf fescues, bentgrasses and bluegrasses.

Pathogen

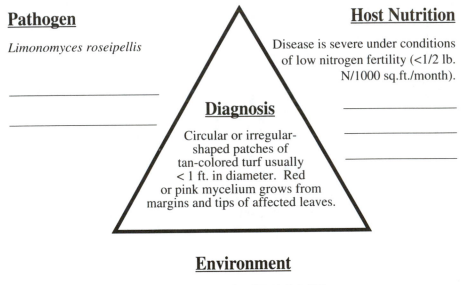

Limonomyces roseipellis

Host Nutrition

Disease is severe under conditions of low nitrogen fertility (<1/2 lb. N/1000 sq.ft./month).

Diagnosis

Circular or irregular-shaped patches of tan-colored turf usually < 1 ft. in diameter. Red or pink mycelium grows from margins and tips of affected leaves.

Environment

Cool temperatures of 60°-75°F (16°-24°C).
More than 10 hrs. of leaf wetness per day for several days.

Pink Patch

Disease Forecasting and Pathogen Detection

A pink patch forecaster and detection kits are not available.

Chemical Controls

Curalan, Prostar, Touché, Vorlan.

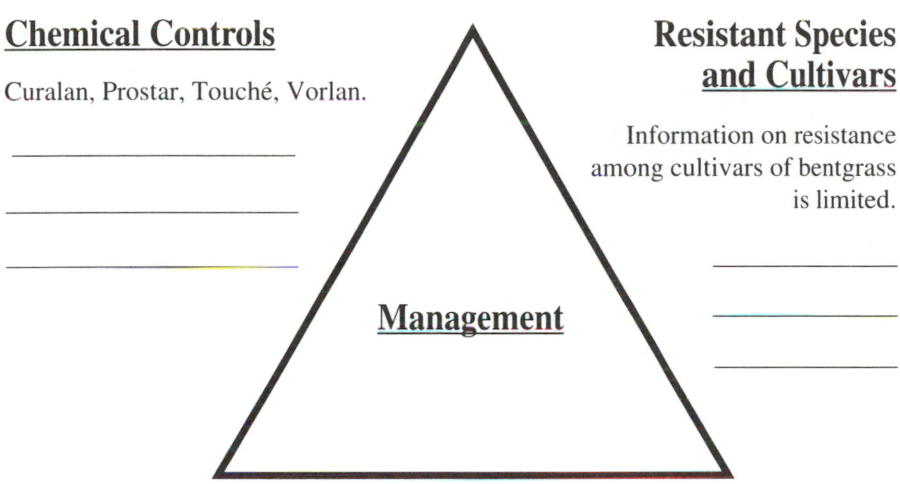

Resistant Species and Cultivars

Information on resistance among cultivars of bentgrass is limited.

Cultural Controls

Avoid low fertility.
Apply at least 1/2 lb. N/1000 sq.ft./month.
Maintain moderate to high levels of phosphorus and potash according to soil tests.
Reduce shade and increase air circulation to enhance drying of turf.
Avoid irrigation in late afternoon or in evening prior to midnight.
Mow turf at least once per week to remove diseased portions of leaf blades.

Pink Snow Mold

Hosts

All cool season turfgrasses. Bentgrasses, annual bluegrass and perennial ryegrass are particularly susceptible.

Pathogen

Microdochium nivale (same pathogen causes Fusarium Patch)

Host Nutrition

Fast-growing, lush turf, that receives high nitrogen and low potash in late fall is particularly susceptible.

Diagnosis

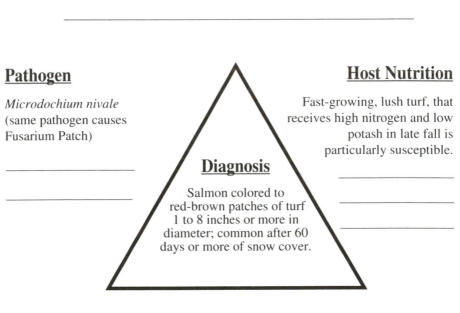

Salmon colored to red-brown patches of turf 1 to 8 inches or more in diameter; common after 60 days or more of snow cover.

Environment

Disease is common after at least 60 days of snow cover, but pathogen can infect turf in absence of snow (see Fusarium Patch).
Disease is particularly severe when snow covers unfrozen ground.

Pink Snow Mold

Disease Forecasting and Pathogen Detection

A pink snow mold forecaster and detection kits are not available.

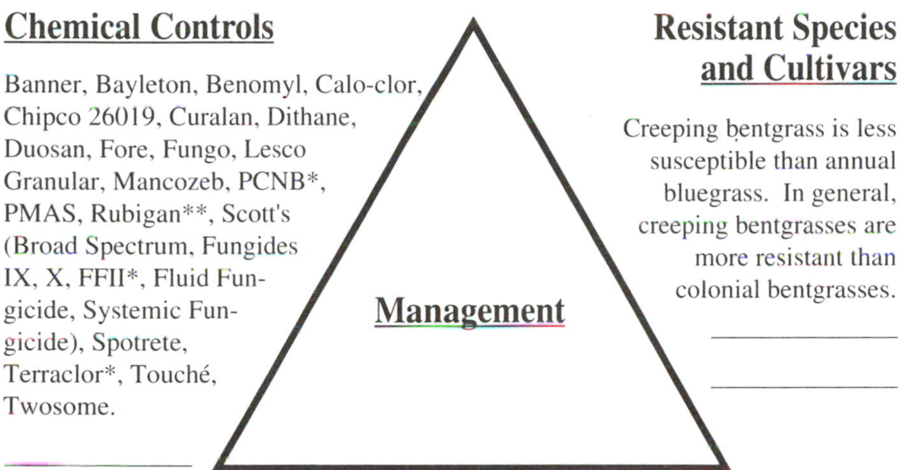

Chemical Controls

Banner, Bayleton, Benomyl, Calo-clor, Chipco 26019, Curalan, Dithane, Duosan, Fore, Fungo, Lesco Granular, Mancozeb, PCNB*, PMAS, Rubigan**, Scott's (Broad Spectrum, Fungides IX, X, FFII*, Fluid Fungicide, Systemic Fungicide), Spotrete, Terraclor*, Touché, Twosome.

Resistant Species and Cultivars

Creeping bentgrass is less susceptible than annual bluegrass. In general, creeping bentgrasses are more resistant than colonial bentgrasses.

Cultural Controls

Maintain moderate nitrogen fertility (1/2 lb. N/1000 sq.ft./month) during late summer and fall.
Maintain high potash levels according to soil tests.
Use snow fence, shrubs or knolls as wind-breaks to prevent excess snow from accumulating.
Prevent snow compaction by machinery or skiers.
Melt snow in spring with organic fertilizers.
Physically remove snow in spring.
Follow controls for Fusarium Patch after snow melt.

* Avoid application to actively growing bentgrasses. ** May reduce populations of annual bluegrass.

Pythium Blight

Hosts

All cool season turfgrasses. Annual bluegrass and perennial ryegrass are particularly susceptible.

Pathogen

Pythium aphanidermatum and other species of *Pythium.*

Host Nutrition

Disease is severe under high nitrogen fertility (> 1/2 lb. N/1000 sq.ft./month). Deficiency in calcium may increase susceptibility.

Diagnosis

Greasy brown patches of turf an inch or less in diameter, increasing to several inches and turning straw colored. Grey-white, cottony mycelium observed in early morning.

Environment

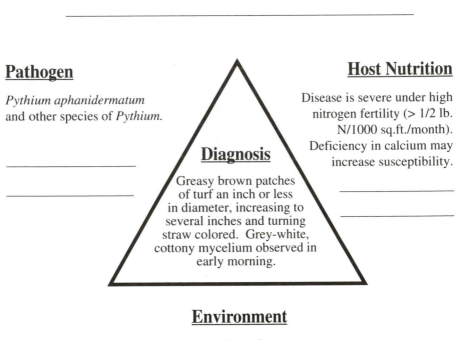

Night temperature > 65°F (18°C).
More than 10 hrs. of leaf wetness per day for several days.
Poor surface and sub-surface drainage.

Pythium Blight

Disease Forecasting and Pathogen Detection

Pythium blight forecasters are available from Pest Management Supply, P.O. Box 936, Amherst, MA 01004 or Neogen Corp., 620 Lesher Pl., Lansing, MI 48912. Detection kits are available from Neogen Corp.

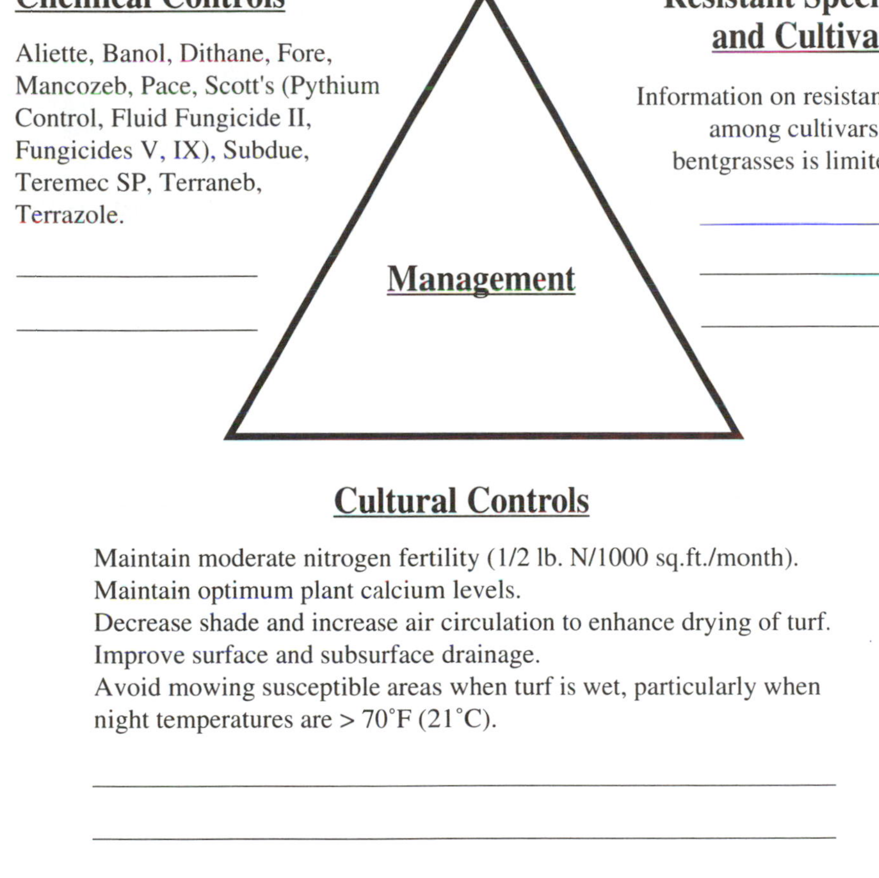

Chemical Controls

Aliette, Banol, Dithane, Fore, Mancozeb, Pace, Scott's (Pythium Control, Fluid Fungicide II, Fungicides V, IX), Subdue, Teremec SP, Terraneb, Terrazole.

Resistant Species and Cultivars

Information on resistance among cultivars of bentgrasses is limited.

Cultural Controls

Maintain moderate nitrogen fertility (1/2 lb. N/1000 sq.ft./month).
Maintain optimum plant calcium levels.
Decrease shade and increase air circulation to enhance drying of turf.
Improve surface and subsurface drainage.
Avoid mowing susceptible areas when turf is wet, particularly when night temperatures are > 70°F (21°C).

Pythium Root Rot

Hosts

All species of cool season grasses.

Pathogen

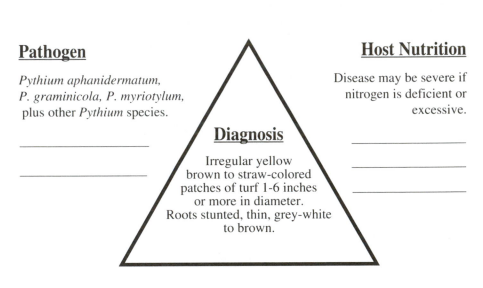

Pythium aphanidermatum, P. graminicola, P. myriotylum, plus other *Pythium* species.

Host Nutrition

Disease may be severe if nitrogen is deficient or excessive.

Diagnosis

Irregular yellow brown to straw-colored patches of turf 1-6 inches or more in diameter. Roots stunted, thin, grey-white to brown.

Environment

Cool (32°-50°F, 0°-10°C) or warm (70°-90°F, 21°-32°C) soil temperatures*.
High soil moisture.
Poor surface or subsurface drainage.
Conditions unfavorable for carbohydrate development by leaves – low light, low mowing height, excessive wear.

* Some *Pythium* species are favored by cool soils, other species by warm soils.

Pythium Root Rot

Disease Forecasting and Pathogen Detection

A Pythium root rot forecaster is not available. Detection kits are available from Neogen Corp., 620 Lesher Pl., Lansing, MI 48912.

Chemical Controls

Banol, Koban, Subdue, Truban.

Resistant Species and Cultivars

Information on resistance among cultivars of bentgrasses is limited.

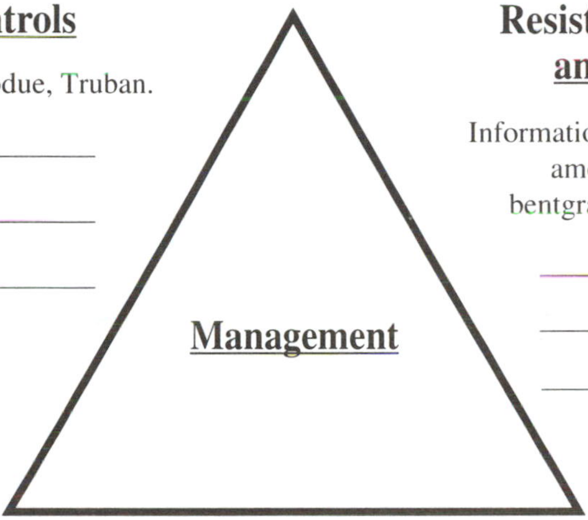

Management

Cultural Controls

Maintain moderate levels of nitrogen (1/2 lb. N/1000 sq.ft./month).
Do not over fertilize with nitrogen in spring when roots are forming.
Maintain moderate to high levels of phosphorus and potash according to soil tests.
Improve surface and subsurface drainage.
Raise mowing height.
Decrease shade.
Use light-weight mowing equipment.
Applications of certain composts may reduce disease severity.

Red Leaf Spot

Hosts

Bentgrasses are susceptible. Disease has not been reported on annual bluegrass.

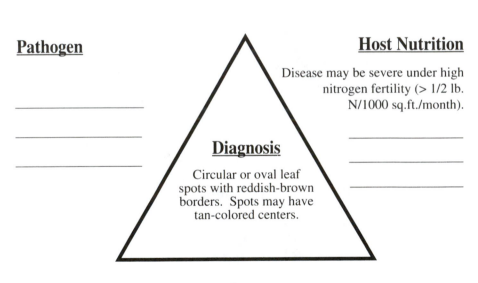

Pathogen

Host Nutrition

Disease may be severe under high nitrogen fertility (> 1/2 lb. N/1000 sq.ft./month).

Diagnosis

Circular or oval leaf spots with reddish-brown borders. Spots may have tan-colored centers.

Environment

Temperature 70°-95°F (21°-35°C).
Disease is particularly severe at high temperatures (>85°F).
More than 10 hrs. of leaf wetness per day for several days.

Red Leaf Spot

Disease Forecasting and Pathogen Detection

A red leaf spot forecaster and detection kits are not available.

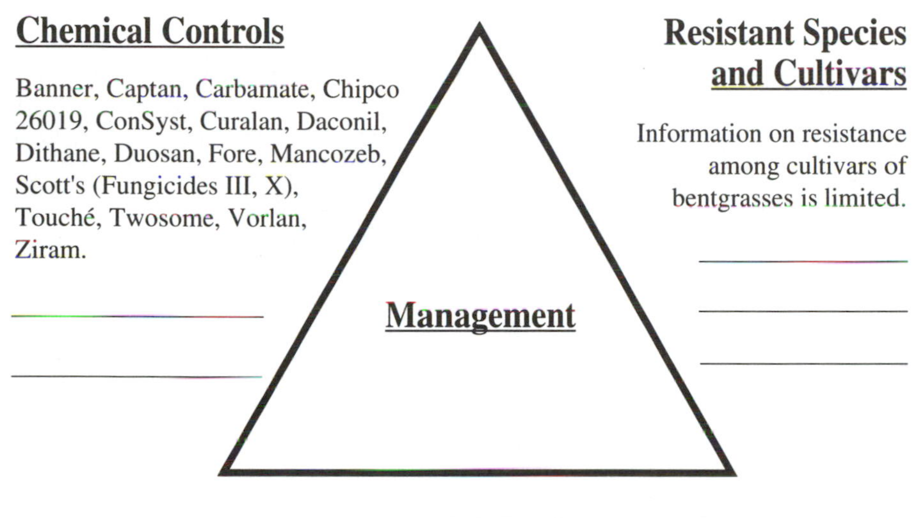

Chemical Controls

Banner, Captan, Carbamate, Chipco 26019, ConSyst, Curalan, Daconil, Dithane, Duosan, Fore, Mancozeb, Scott's (Fungicides III, X), Touché, Twosome, Vorlan, Ziram.

Resistant Species and Cultivars

Information on resistance among cultivars of bentgrasses is limited.

Cultural Controls

Maintain moderate nitrogen fertility (1/2 lb. N/1000 sq.ft./month).
Maintain moderate to high levels of phosphorus and potash according to soil tests.
Decrease shade and increase air circulation to enhance drying of turf.
Avoid irrigation in late afternoon and in evening prior to midnight.
Raise mowing height.
Use light-weight mowing equipment to avoid stress on turf.
Limit thatch thickness to 1/4 inch or less.

Red Thread

Hosts

Bentgrasses, bluegrasses, fine-leaf fescues and perennial ryegrass. Fine-leaf fescues and perennial ryegrass are particularly susceptible.

Pathogen

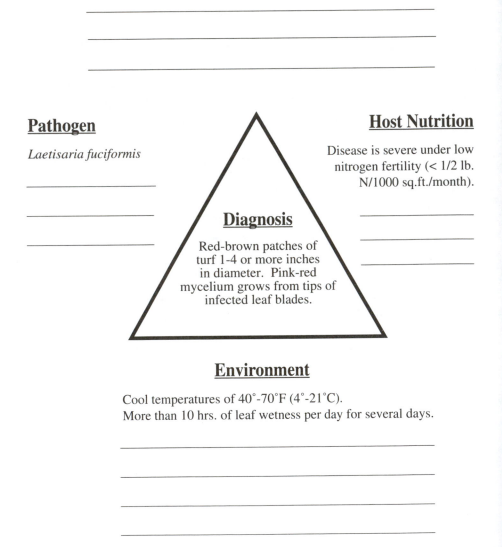

Laetisaria fuciformis

Host Nutrition

Disease is severe under low nitrogen fertility (< 1/2 lb. N/1000 sq.ft./month).

Diagnosis

Red-brown patches of turf 1-4 or more inches in diameter. Pink-red mycelium grows from tips of infected leaf blades.

Environment

Cool temperatures of 40°-70°F (4°-21°C).
More than 10 hrs. of leaf wetness per day for several days.

Red Thread

Disease Forecasting and Pathogen Detection

A red thread forecaster and detection kits are not available.

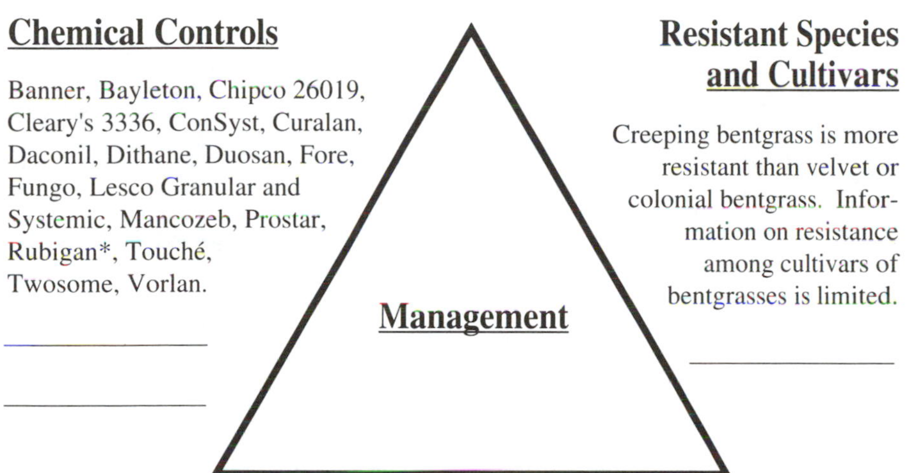

Chemical Controls

Banner, Bayleton, Chipco 26019, Cleary's 3336, ConSyst, Curalan, Daconil, Dithane, Duosan, Fore, Fungo, Lesco Granular and Systemic, Mancozeb, Prostar, Rubigan*, Touché, Twosome, Vorlan.

Resistant Species and Cultivars

Creeping bentgrass is more resistant than velvet or colonial bentgrass. Information on resistance among cultivars of bentgrasses is limited.

Cultural Controls

Fertilize with at least 1/2 lb. N/1000 sq.ft./month.
Maintain moderate to high levels of potash and phosphorus according to soil tests.
Reduce shade and increase air circulation to enhance drying of turf.
Avoid irrigation in late afternoon or in evening prior to midnight.
Maintian soil pH at 6.5 to 7.0.
Mow turf at least once per week to remove diseased portion of leaf blades.

* May reduce populations of annual bluegrass.

Rhizoctonia Leaf and Sheath Spot

Hosts

Bentgrasses, bluegrasses, perennial ryegrass and tall fescue.

Pathogen

Rhizoctonia zeae

Host Nutrition

Effects of nutrition on disease development are unknown.

Diagnosis

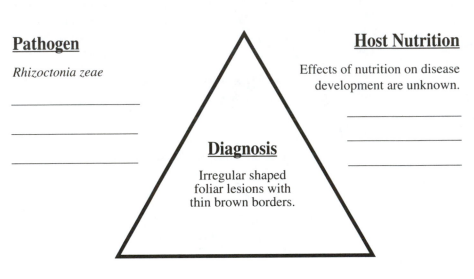

Irregular shaped foliar lesions with thin brown borders.

Environment

Day temperatures > 90°F (32°C).
More than 10 hrs. of leaf wetness per day for several days.

Rhizoctonia Leaf and Sheath Spot

Disease Forecasting and Pathogen Detection

A leaf and sheath spot forecaster and detection kits are not available.

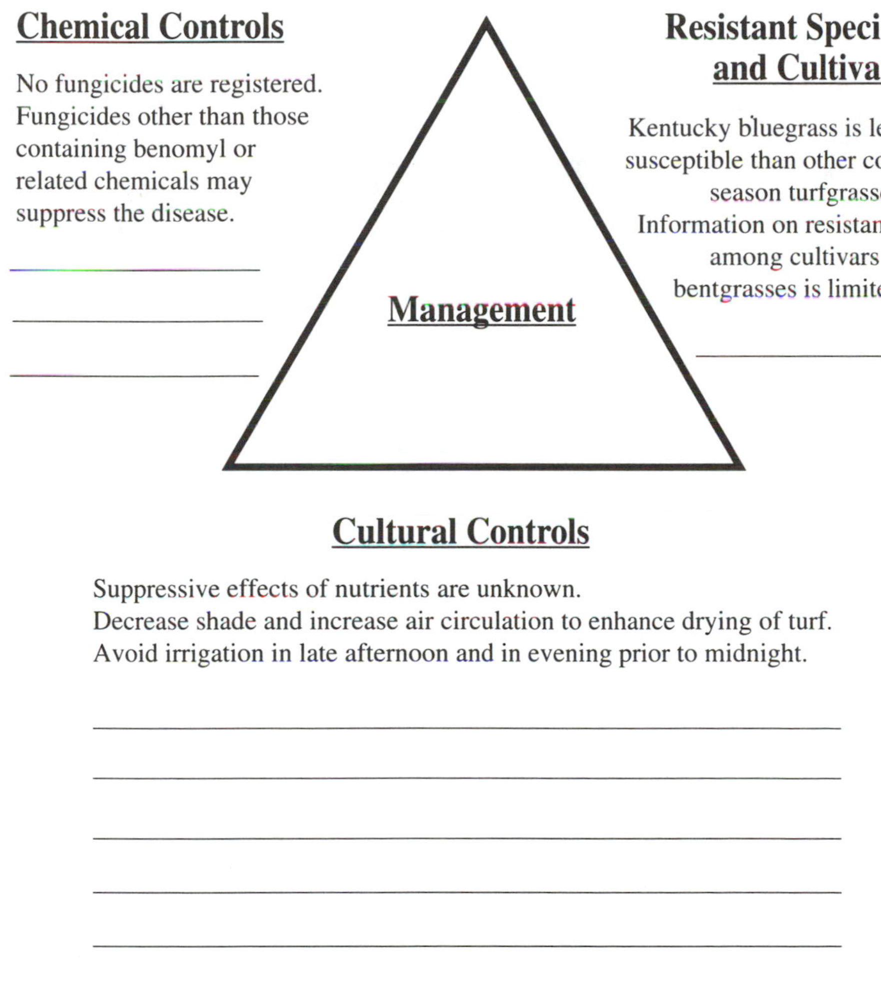

Chemical Controls

No fungicides are registered. Fungicides other than those containing benomyl or related chemicals may suppress the disease.

Resistant Species and Cultivars

Kentucky bluegrass is less susceptible than other cool season turfgrasses. Information on resistance among cultivars of bentgrasses is limited.

Management

Cultural Controls

Suppressive effects of nutrients are unknown.
Decrease shade and increase air circulation to enhance drying of turf.
Avoid irrigation in late afternoon and in evening prior to midnight.

Rust

Hosts

All common species of cool season turfgrasses.

Pathogen

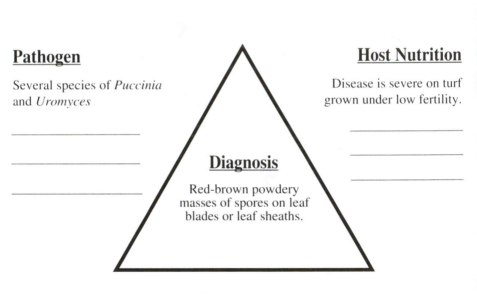

Several species of *Puccinia* and *Uromyces*

Host Nutrition

Disease is severe on turf grown under low fertility.

Diagnosis

Red-brown powdery masses of spores on leaf blades or leaf sheaths.

Environment

Temperatures of 68°- 86°F (20°- 30°C).
Disease is severe on turf subjected to drought stress, low mowing, shade or poor air circulation.

Rust

Disease Forecasting and Pathogen Detection

A rust forecaster and detection kits are not available.

Chemical Controls

Banner, Bayleton, Captan, Carbamate, Cleary's 3336, ConSyst, Daconil, Dithane, Duosan, Fore, Mancozeb, Rubigan**, Scott's (Fluid Fungicide III, Fungicide VII, FFII*), Twosome, Ziram.

Resistant Species and Cultivars

Information on resistance among cultivars of bentgrass is limited.

Management

Cultural Controls

Maintain moderate and balanced fertility throughout the growing season.
Reduce shade and increase air circulation.
Increase mowing height.
Avoid drought stress.
Avoid irrigation in late afternoon and in evening prior to midnight.

* Avoid application to bentgrasses. ** May reduce populations of annual bluegrass.

Sclerotium Blight
(Southern Blight)

Hosts

Bentgrasses and annual bluegrass.

Pathogen

Sclerotium rolfsii

Host Nutrition

Little information is available on effect of fertility on Sclerotium blight.

Diagnosis

Yellow to white patches of turf a few inches to several feet in diameter. Patches have red-brown borders with white mycelium and white to brown sclerotia near the soil surface.

Environment

High day-time temperature >85°F (29°C).
Night temperatures >70°F (21°C).
Disease is severe when wet weather follows a period of drought.
Disease may be severe at soil pH of 6.5 or less.

Sclerotium Blight

Disease Forecasting and Pathogen Detection

A Sclerotium blight forecaster and detection kits are not available.

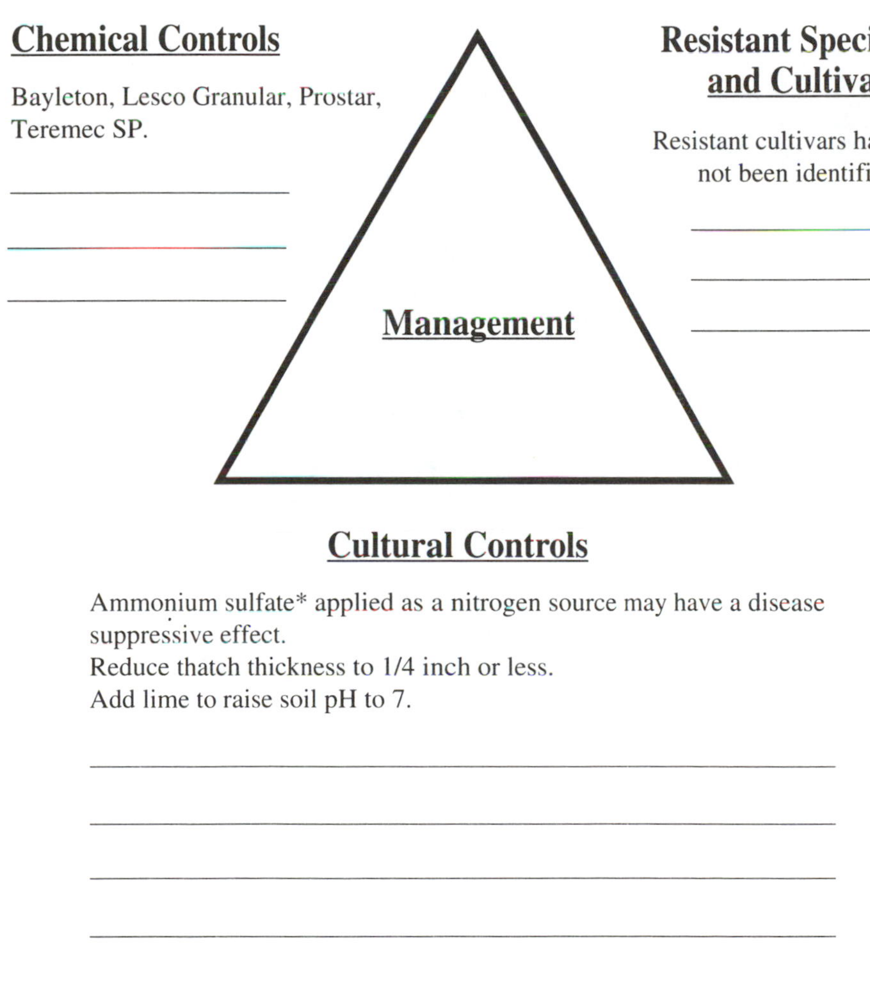

Chemical Controls

Bayleton, Lesco Granular, Prostar, Teremec SP.

Resistant Species and Cultivars

Resistant cultivars have not been identified.

Cultural Controls

Ammonium sulfate* applied as a nitrogen source may have a disease suppressive effect.
Reduce thatch thickness to 1/4 inch or less.
Add lime to raise soil pH to 7.

* Summer applications of ammonium sulfate should be avoided because of high potential for foliar burn.

Stripe Smut

Hosts

Bluegrasses, bentgrasses, perennial ryegrass and tall fescue. Kentucky bluegrass is more susceptible than other cool season grasses.

Pathogen

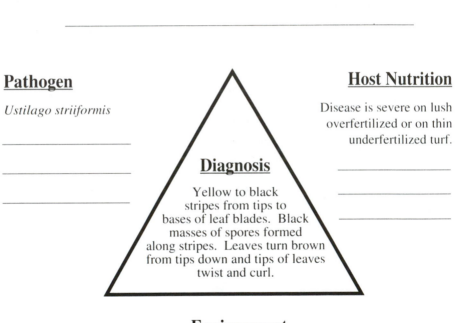

Ustilago striiformis

Host Nutrition

Disease is severe on lush overfertilized or on thin underfertilized turf.

Diagnosis

Yellow to black stripes from tips to bases of leaf blades. Black masses of spores formed along stripes. Leaves turn brown from tips down and tips of leaves twist and curl.

Environment

Infection occurs at 50°- 68°F (10°- 20°C).
Severe symptoms are evident during drought and temperatures >75°F (24°C).
Symptoms are often more severe on acid soils and on turf with excessive thatch (>1/2 inch thick).

Stripe Smut

Disease Forecasting and Pathogen Detection

A stripe smut forecaster and detection kits are not available.

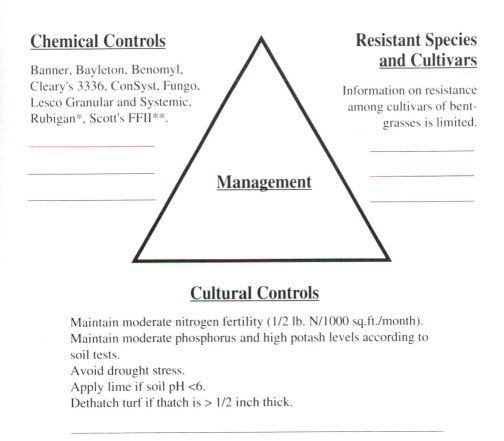

Chemical Controls

Banner, Bayleton, Benomyl, Cleary's 3336, ConSyst, Fungo, Lesco Granular and Systemic, Rubigan*, Scott's FFII**.

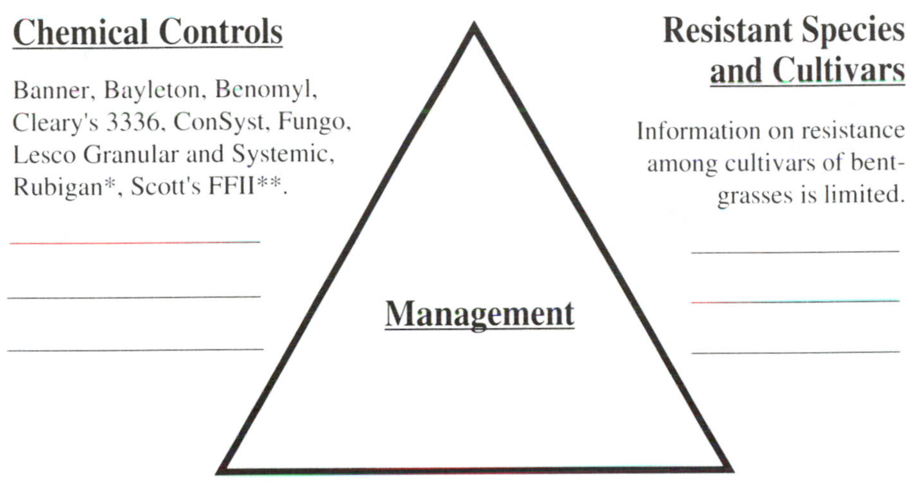

Resistant Species and Cultivars

Information on resistance among cultivars of bentgrasses is limited.

Cultural Controls

Maintain moderate nitrogen fertility (1/2 lb. N/1000 sq.ft./month).
Maintain moderate phosphorus and high potash levels according to soil tests.
Avoid drought stress.
Apply lime if soil pH <6.
Dethatch turf if thatch is > 1/2 inch thick.

* May reduce population of annual bluegrass. ** Avoid application to bentgrasses.

Summer Patch

Hosts

Bluegrasses and fine-leaf fescues.

Pathogen

Magnaporthe poae

Host Nutrition

Disease may be severe when turf is fertilized with fast-release sources of nitrogen.

Diagnosis

Circular patches of wilted to straw-colored turf, usually less than 10 inches in diameter. Leaves turn yellow or brown starting at tips. Roots are light to dark brown.

Environment

Day-time temperature >85°F (29°C).
High soil moisture.
Poor surface or subsurface drainage.
Low mowing height.

Summer Patch

Disease Forecasting and Pathogen Detection

A summer patch forecaster and detection kits are not available.

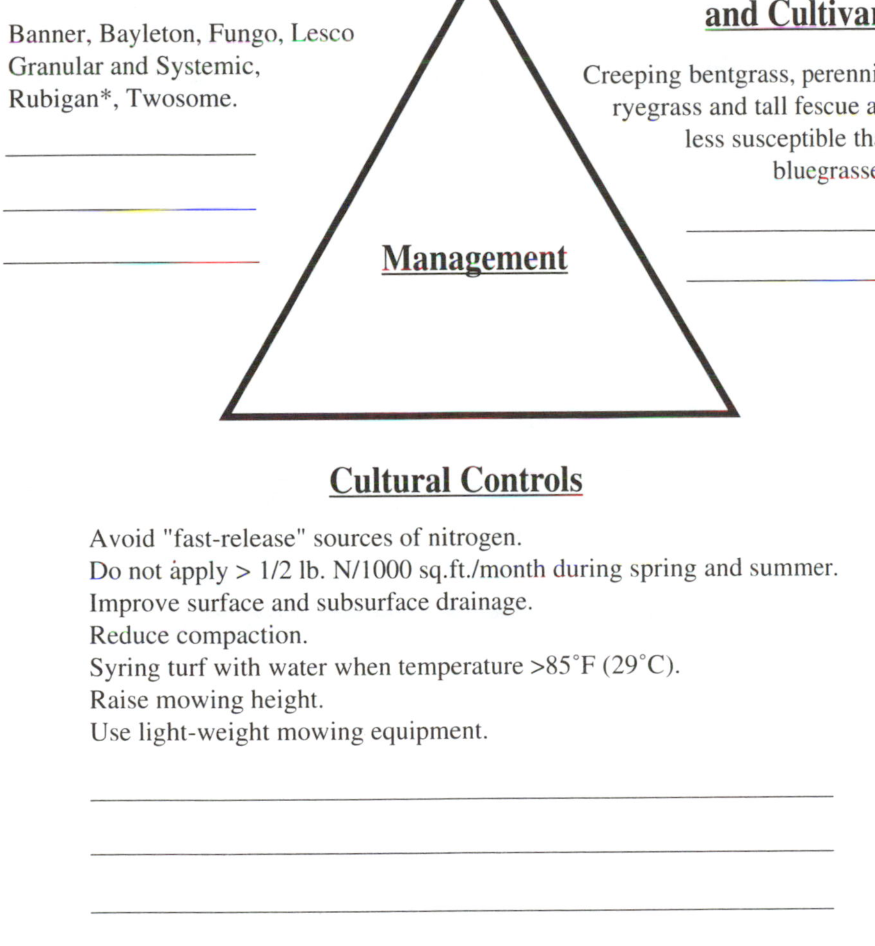

Chemical Controls

Banner, Bayleton, Fungo, Lesco Granular and Systemic, Rubigan*, Twosome.

Resistant Species and Cultivars

Creeping bentgrass, perennial ryegrass and tall fescue are less susceptible than bluegrasses.

Cultural Controls

Avoid "fast-release" sources of nitrogen.
Do not apply > 1/2 lb. N/1000 sq.ft./month during spring and summer.
Improve surface and subsurface drainage.
Reduce compaction.
Syring turf with water when temperature >85˚F (29˚C).
Raise mowing height.
Use light-weight mowing equipment.

* May reduce populations of annual bluegrass.

Take-All Patch

Hosts

Only bentgrasses are highly susceptible.

Pathogen

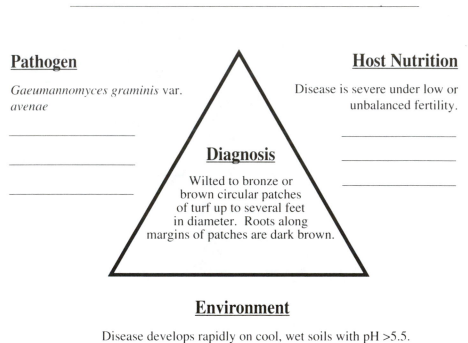

Gaeumannomyces graminis var. *avenae*

Host Nutrition

Disease is severe under low or unbalanced fertility.

Diagnosis

Wilted to bronze or brown circular patches of turf up to several feet in diameter. Roots along margins of patches are dark brown.

Environment

Disease develops rapidly on cool, wet soils with pH >5.5. Disease can be severe on sandy soils.

Take-All Patch

Disease Forecasting and Pathogen Detection

A take-all patch forecaster and detection kits are not available.

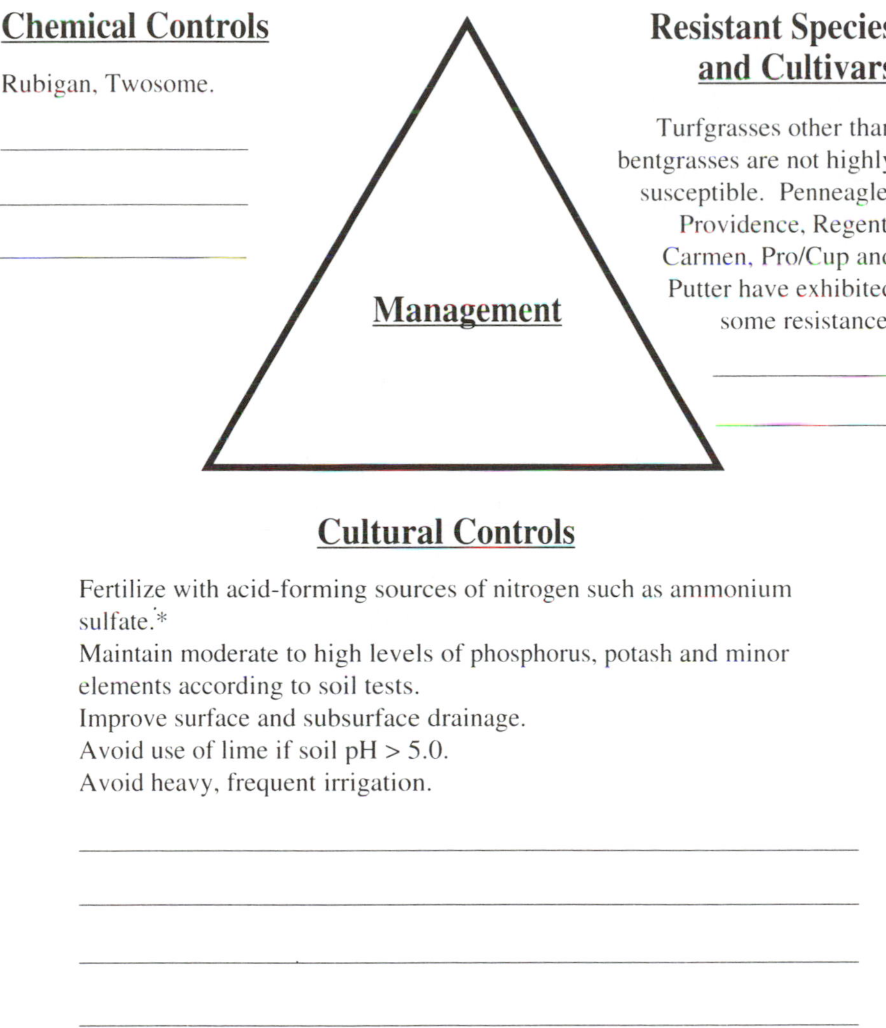

Chemical Controls

Rubigan, Twosome.

Resistant Species and Cultivars

Turfgrasses other than bentgrasses are not highly susceptible. Penneagle, Providence, Regent, Carmen, Pro/Cup and Putter have exhibited some resistance.

Management

Cultural Controls

Fertilize with acid-forming sources of nitrogen such as ammonium sulfate.*
Maintain moderate to high levels of phosphorus, potash and minor elements according to soil tests.
Improve surface and subsurface drainage.
Avoid use of lime if soil pH > 5.0.
Avoid heavy, frequent irrigation.

* Summer applications of ammonium sulfate should be avoided because of high potential for foliar burn.

Typhula Blight
(Gray Snow Mold)

Hosts

All cool season turfgrasses. Bentgrasses, annual bluegrass and perennial ryegrass are particularly susceptible.

Pathogen

Typhula incarnata and *T. ishikariensis*

Host Nutrition

Disease is severe on lush, fast-growing turf that is covered with snow. High nitrogen and low potash in fall can predispose turf to severe damage.

Diagnosis

Circular straw-colored patches of turf usually less than 10 inches in diameter, evident after snow-melt. Orange, brown to black sclerotia form on leaves.

Environment

Snow-cover is required for disease development. Disease is severe when snow-cover exceeds 90 days.

Typhula Blight

Disease Forecasting and Pathogen Detection

A Typhula blight forecaster and detection kits are not available.

Chemical Controls

Banner, Bayleton, Calo-clor, Calo-gran, Chipco 26019, Curalan, Daconil, Lesco Granular, PCNB*, PMAS, Prostar, Rubigan**, Scott's (Broad Spectrum, Fungicides V, IX, FFII*), Spotrete, Terraclor*, Teremec SP, Thiram, Touché, Turfcide*, Twosome.

Resistant Species and Cultivars

No cultivars of creeping bentgrass or annual bluegrass are reported to be resistant. Browntop bentgrasses are more resistant than creeping bentgrasses.

Management

Cultural Controls

Avoid a fertility program that results in lush, fast-growing turf in late fall and winter.
Maintain high potash levels according to soil tests.
Use snow fence, hedges or knolls to prevent snow from accumulating excessively on turf.
Use dark-colored organic fertilizers or composts to melt snow in spring.
Physically remove snow in spring.
Prevent compaction of snow during winter.

* Avoid application to actively growing creeping bentgrass. ** May reduce populations of annual bluegrass.

White Patch

(White Blight)

Hosts

Fescues, bluegrasses and creeping bentgrass.

Pathogen

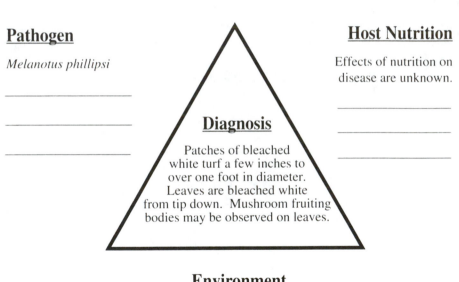

Melanotus phillipsi

Host Nutrition

Effects of nutrition on disease are unknown.

Diagnosis

Patches of bleached white turf a few inches to over one foot in diameter. Leaves are bleached white from tip down. Mushroom fruiting bodies may be observed on leaves.

Environment

Night temperatures >70°F (21°C).
More than 10 hrs. of leaf wetness per day for several days.
Disease is particularly severe on soils from recently cleared forests.

White Patch

Disease Forecasting and Pathogen Detection

A white patch forecaster and detection kits are not available.

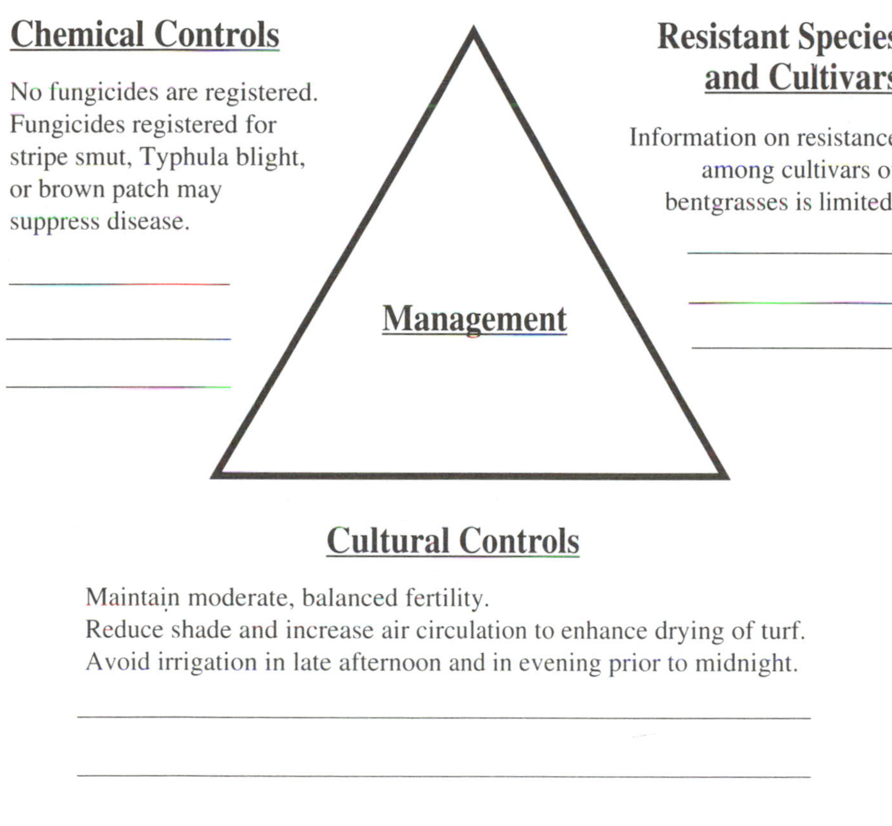

Chemical Controls

No fungicides are registered. Fungicides registered for stripe smut, Typhula blight, or brown patch may suppress disease.

Resistant Species and Cultivars

Information on resistance among cultivars of bentgrasses is limited.

Management

Cultural Controls

Maintain moderate, balanced fertility.
Reduce shade and increase air circulation to enhance drying of turf.
Avoid irrigation in late afternoon and in evening prior to midnight.

Yellow Patch

Hosts

Bentgrasses, bluegrasses and tall fescue.

Pathogen

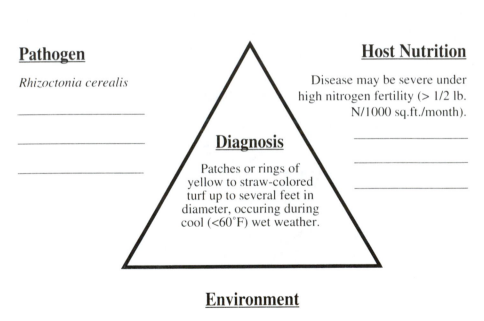

Rhizoctonia cerealis

Host Nutrition

Disease may be severe under high nitrogen fertility (> 1/2 lb. N/1000 sq.ft./month).

Diagnosis

Patches or rings of yellow to straw-colored turf up to several feet in diameter, occuring during cool (<60°F) wet weather.

Environment

Temperatures <60°F (16°C).
More than 10 hrs. of leaf wetness per day for several days.
Disease is severe on turf with excessive thatch.

Yellow Patch

Disease Forecasting and Pathogen Detection

A yellow patch forecaster and detection kits are not available.

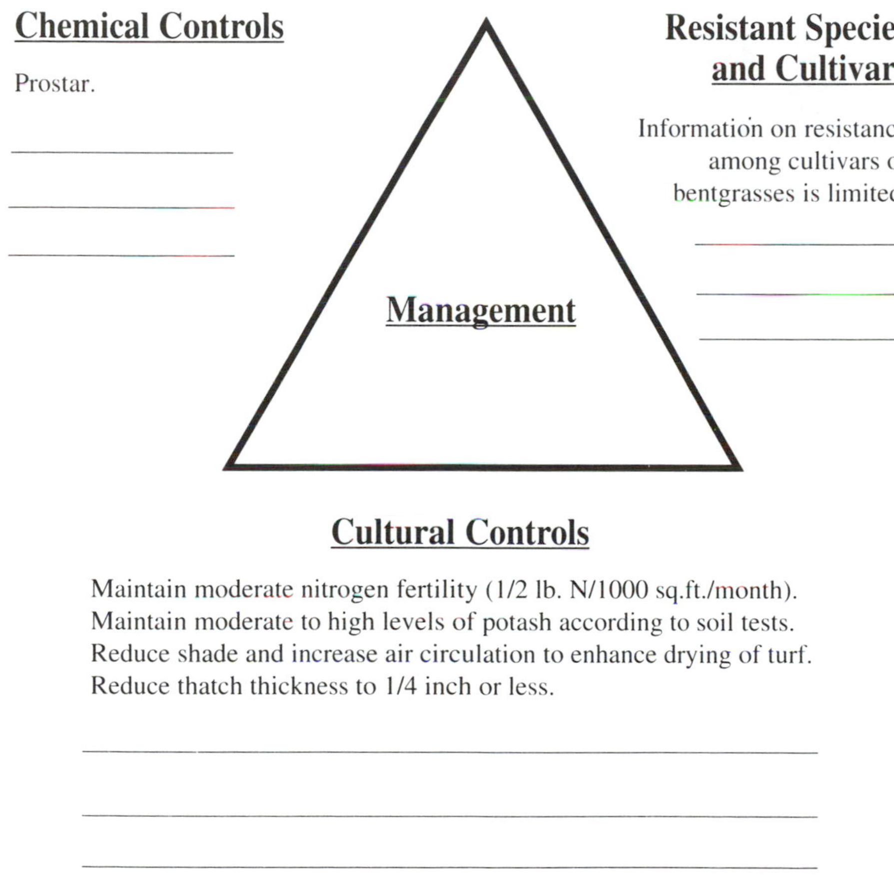

Chemical Controls

Prostar.

Resistant Species and Cultivars

Information on resistance among cultivars of bentgrasses is limited.

Cultural Controls

Maintain moderate nitrogen fertility (1/2 lb. N/1000 sq.ft./month).
Maintain moderate to high levels of potash according to soil tests.
Reduce shade and increase air circulation to enhance drying of turf.
Reduce thatch thickness to 1/4 inch or less.

Yellow Ring

Hosts

Bluegrasses and bentgrasses.

Pathogen

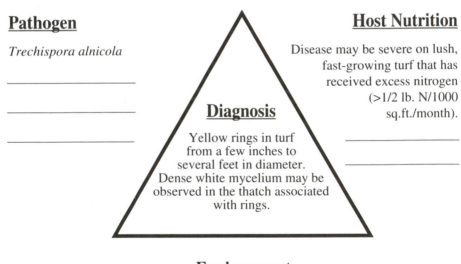

Trechispora alnicola

Host Nutrition

Disease may be severe on lush, fast-growing turf that has received excess nitrogen (>1/2 lb. N/1000 sq.ft./month).

Diagnosis

Yellow rings in turf from a few inches to several feet in diameter. Dense white mycelium may be observed in the thatch associated with rings.

Environment

Temperatures between 60°- 90°F (15°-32°C).
Disease is severe on turf with excessive thatch (> 1/2 inch thick).

Yellow Ring

Disease Forecasting and Pathogen Detection

A yellow ring forecaster and detection kits are not available.

Chemical Controls

No fungicides are registered. Fungicides containing PCNB have suppressed disease in test plots.

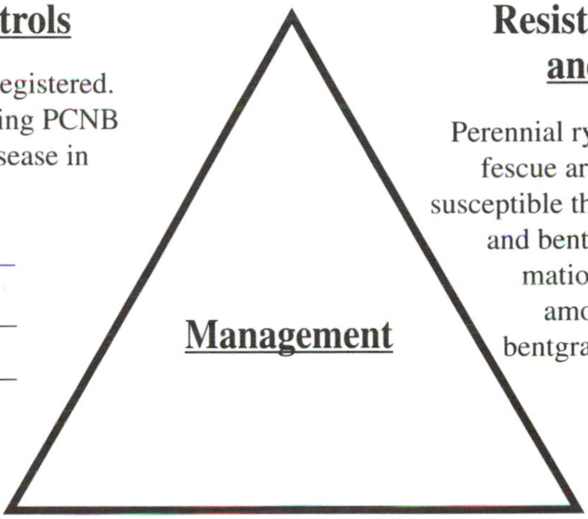

Resistant Species and Cultivars

Perennial ryegrass and tall fescue are probably less susceptible than bluegrasses and bentgrasses. Information on resistance among cultivars of bentgrasses is limited.

Management

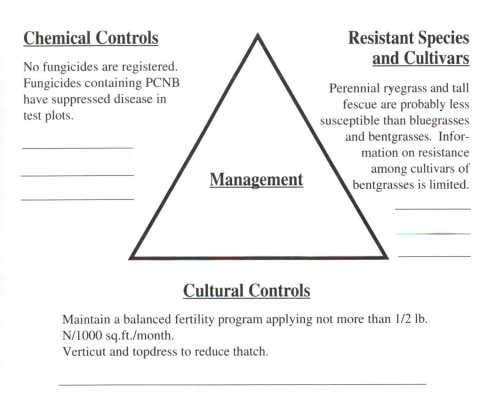

Cultural Controls

Maintain a balanced fertility program applying not more than 1/2 lb. N/1000 sq.ft./month.
Verticut and topdress to reduce thatch.

Yellow Tuft

Hosts

Bentgrasses and annual bluegrass.

Pathogen

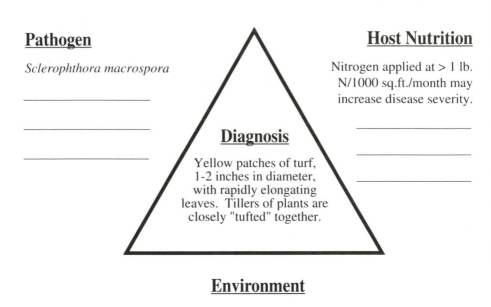

Sclerophthora macrospora

Host Nutrition

Nitrogen applied at > 1 lb. N/1000 sq.ft./month may increase disease severity.

Diagnosis

Yellow patches of turf, 1-2 inches in diameter, with rapidly elongating leaves. Tillers of plants are closely "tufted" together.

Environment

Temperatures from 40°-70°F (5°-20°C).
More than 10 hrs. of leaf wetness per day for several days.
Poor drainage.

Yellow Tuft

Disease Forecasting and Pathogen Detection

A yellow tuft forecaster and detection kits are not available.

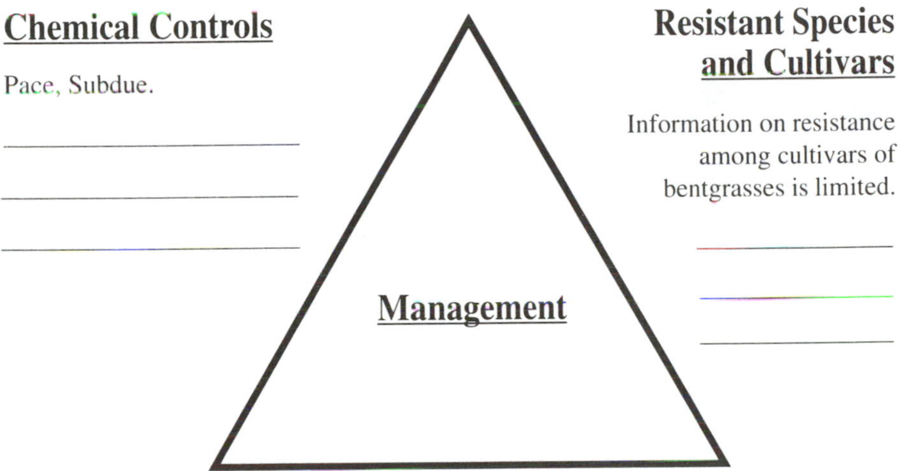

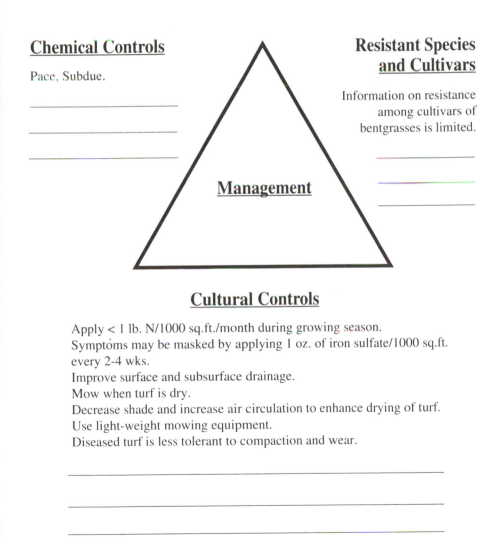

Chemical Controls

Pace, Subdue.

Resistant Species and Cultivars

Information on resistance among cultivars of bentgrasses is limited.

Cultural Controls

Apply < 1 lb. N/1000 sq.ft./month during growing season.
Symptoms may be masked by applying 1 oz. of iron sulfate/1000 sq.ft. every 2-4 wks.
Improve surface and subsurface drainage.
Mow when turf is dry.
Decrease shade and increase air circulation to enhance drying of turf.
Use light-weight mowing equipment.
Diseased turf is less tolerant to compaction and wear.

Chapter 2

Kentucky Bluegrass

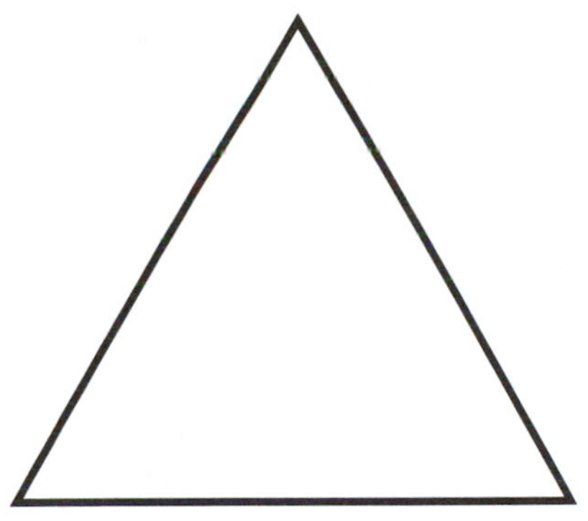

Brown Patch
(Rhizoctonia Blight)

Hosts

All common species of turfgrasses.

Pathogen

Rhizoctonia solani

Host Nutrition

Disease is severe on lush turf fertilized with excessive nitrogen (> 1/2 lb. N/1000 sq.ft./month). Disease is severe on soils low in phosphorus and potash.

Diagnosis

Circular patches of brown turf a few inches to several feet in diameter. Leaves at margins of the patches have gray irregular-shaped lesions with thin brown borders.

Environment

Night temperatures > 60°F (16°C).
More than 10 hrs. of foliar wetness per day for several days.
Disease is severe at low mowing heights (<2 inches).

Brown Patch

(Rhizoctonia Blight)

Disease Forecasting and Pathogen Detection

A brown patch forecaster and detection kits are available from Neogen Corp., 620 Lesher Pl., Lansing, MI 48912.

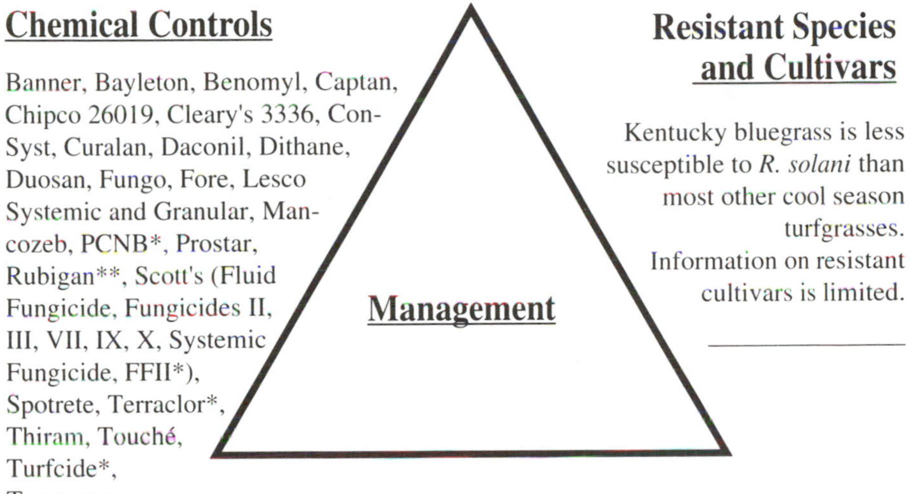

Chemical Controls

Banner, Bayleton, Benomyl, Captan, Chipco 26019, Cleary's 3336, Con-Syst, Curalan, Daconil, Dithane, Duosan, Fungo, Fore, Lesco Systemic and Granular, Mancozeb, PCNB*, Prostar, Rubigan**, Scott's (Fluid Fungicide, Fungicides II, III, VII, IX, X, Systemic Fungicide, FFII*), Spotrete, Terraclor*, Thiram, Touché, Turfcide*, Twosome.

Resistant Species and Cultivars

Kentucky bluegrass is less susceptible to *R. solani* than most other cool season turfgrasses. Information on resistant cultivars is limited.

Cultural Controls

Maintain moderate nitrogen fertility (1/2 lb. N/1000 sq.ft./month).
Maintain moderate phosphorous and high potash according to soil tests.
Decrease shade and increase air circulation to enhance drying of turf.
Avoid irrigation in late afternoon and in evening prior to midnight.
Maintain thatch at 1/2 inch thick or less.
Avoid mowing at height of <2 inches.

* Avoid application to bentgrasses. ** May reduce populations of annual bluegrass.

Curvularia Blight

Hosts

Bluegrasses, bentgrasses and perennial ryegrass.

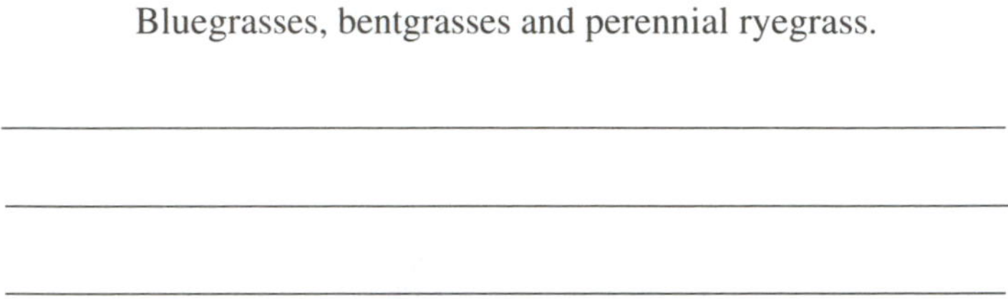

Pathogen

Curvularia species

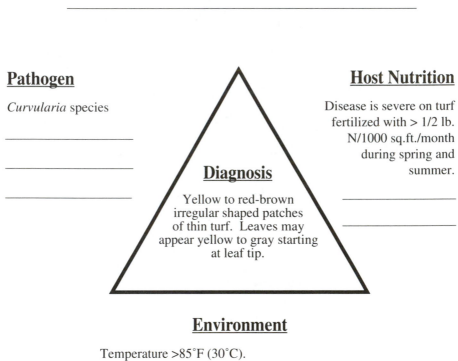

Host Nutrition

Disease is severe on turf fertilized with > 1/2 lb. N/1000 sq.ft./month during spring and summer.

Diagnosis

Yellow to red-brown irregular shaped patches of thin turf. Leaves may appear yellow to gray starting at leaf tip.

Environment

Temperature >85°F (30°C).
More than 10 hrs. of leaf wetness per day for several days.
Thatch > 1/2 inch thick.

Curvularia Blight

Disease Forecasting and Pathogen Detection

A curvularia blight forecaster and detection kits are not available.

Chemical Controls

ConSyst, Daconil, Twosome.

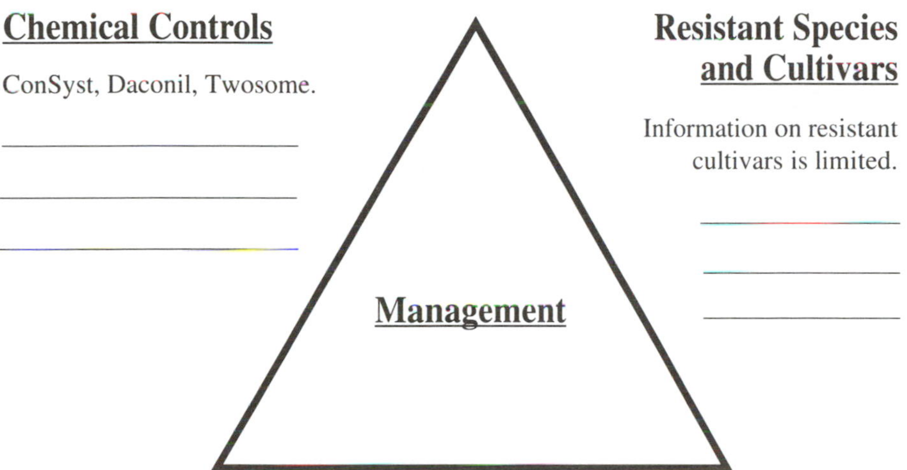

Resistant Species and Cultivars

Information on resistant cultivars is limited.

Cultural Controls

Fertilize with not more than 1/2 lb. N/1000 sq.ft./month during spring and summer.
Use light-weight mowing equipment (reduce compaction).
Limit thatch thickness to 1/2 inch or less.
Decrease shade and increase air circulation to enhance drying of turf.
Syringe turf with water when temperature >80°F (27°C).
Avoid irrigation in late afternoon and in evening prior to midnight.
Avoid mowing at height of < 2 inches.

Damping-off and Seed Rot

Hosts

All species of turfgrasses.

Pathogen

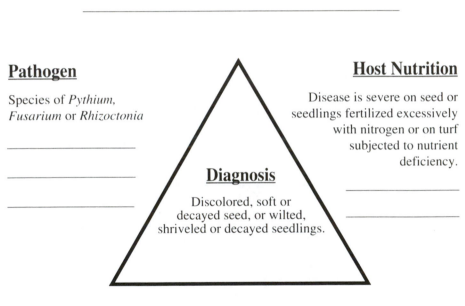

Species of *Pythium, Fusarium* or *Rhizoctonia*

Host Nutrition

Disease is severe on seed or seedlings fertilized excessively with nitrogen or on turf subjected to nutrient deficiency.

Diagnosis

Discolored, soft or decayed seed, or wilted, shriveled or decayed seedlings.

Environment

Temperatures too high (>85˚F, 30˚C) or too low (<60˚F, 15˚C) for optimum seedling development.
More than 10 hrs. of seed or seedling wetness per day for several days.
Excessive shade or overcrowding of seedlings.
Poor surface or subsurface drainage.

Damping-off and Seed Rot

Disease Forecasting and Pathogen Detection

A damping-off or seed rot forecaster is not available. Detection kits for ***Pythium*** and ***Rhizoctonia*** are available from Neogen Corp., 620 Lesher Pl., Lansing, MI 48912.

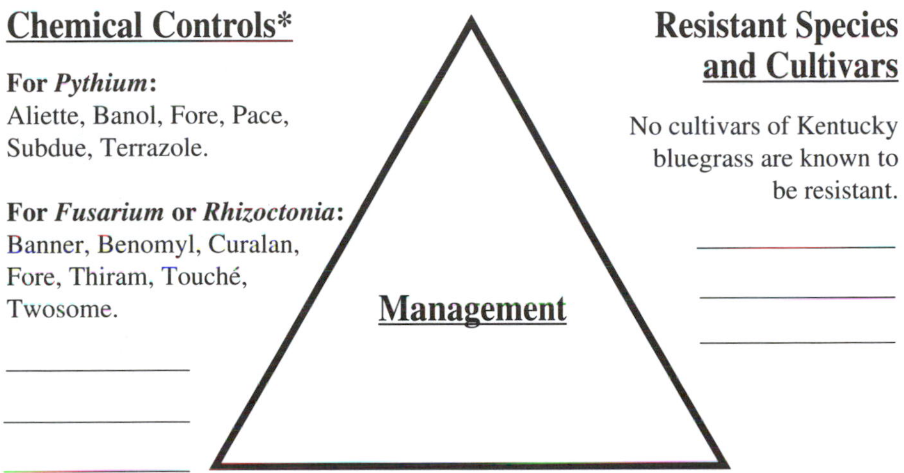

Chemical Controls*

For *Pythium*:
Aliette, Banol, Fore, Pace, Subdue, Terrazole.

For *Fusarium* or *Rhizoctonia*:
Banner, Benomyl, Curalan, Fore, Thiram, Touché, Twosome.

Resistant Species and Cultivars

No cultivars of Kentucky bluegrass are known to be resistant.

Cultural Controls

Incorporate 1-3 lbs. N/1000 sq.ft. in seed bed prior to seeding.
Incorporate phosphorous and potash according to soil tests.
Fertilize seedlings with 1/2 to 1 lb. N/1000 sq.ft./month.
Seed turf when day temperatures are between 60° - 80°F (15° -27°C).
Use recommended seeding rate (1 to 1-1/2 lb. seed/1000 sq.ft.) for Kentucky bluegrass.
Avoid high seeding rates.
Avoid irrigation in late afternoon and in evening prior to midnight.
Reduce shade and increase air circulation to enhance drying of turf.
Avoid mowing at height of < 2 inches.
Improve surface and subsurface drainage.

* Some materials may be sold as seed treatments as well as foliar sprays.

Dollar Spot

Hosts

All common species of turfgrasses.

Pathogen

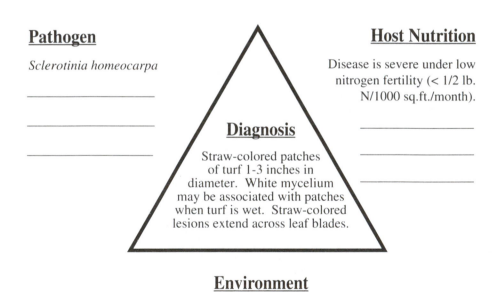

Sclerotinia homeocarpa

Host Nutrition

Disease is severe under low nitrogen fertility (< 1/2 lb. N/1000 sq.ft./month).

Diagnosis

Straw-colored patches of turf 1-3 inches in diameter. White mycelium may be associated with patches when turf is wet. Straw-colored lesions extend across leaf blades.

Environment

Night temperatures >50°F (10°C) and day temperatures <90°F (32°C).
More than 10 hrs. of leaf wetness per day for several days.
Disease is severe on turf subjected to drought stress.

Dollar Spot

Disease Forecasting and Pathogen Detection

A dollar spot forecaster is available from
Pest Management Supply, P.O. Box 938, Amherst, MA 01004.
Detection kits are available from Neogen Corp.,
620 Lesher Pl., Lansing, MI 48912.

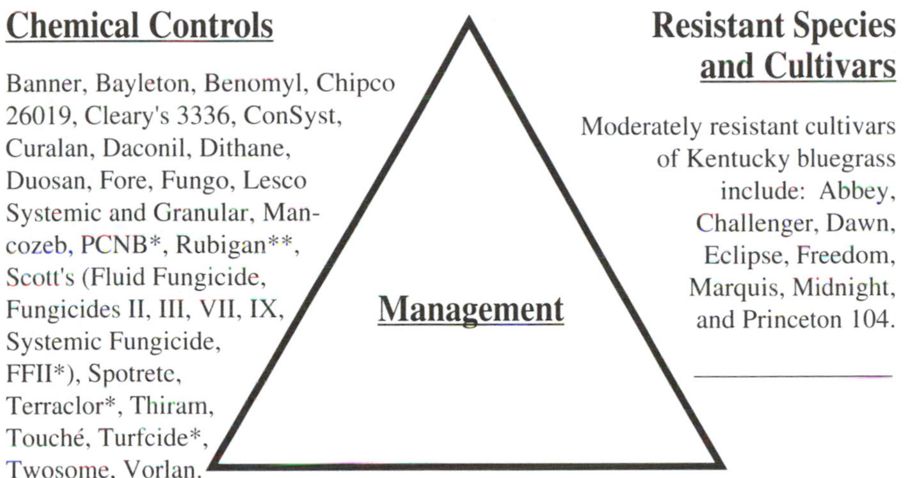

Chemical Controls

Banner, Bayleton, Benomyl, Chipco 26019, Cleary's 3336, ConSyst, Curalan, Daconil, Dithane, Duosan, Fore, Fungo, Lesco Systemic and Granular, Mancozeb, PCNB*, Rubigan**, Scott's (Fluid Fungicide, Fungicides II, III, VII, IX, Systemic Fungicide, FFII*), Spotrete, Terraclor*, Thiram, Touché, Turfcide*, Twosome, Vorlan.

Resistant Species and Cultivars

Moderately resistant cultivars of Kentucky bluegrass include: Abbey, Challenger, Dawn, Eclipse, Freedom, Marquis, Midnight, and Princeton 104.

Cultural Controls

Applications of 1/2 to 1 lb. of N/1000 sq.ft. every 2-4 weeks will reduce severity of dollar spot.
Maintain moderate to high levels of soil potassium as determined by soil tests.
Limit thatch to 1/2 inch or less.
Decrease shade and increase air circulation to enhance drying of turf.
Avoid irrigation in late afternoon and in evening prior to midnight.
Avoid drought stress.
Avoid mowing at height < 2 inches.

* Avoid application to bentgrasses. ** May reduce populations of annual bluegrass.

Fairy Ring

Hosts

All turfgrasses.

Pathogen

Several species of "mushroom-forming" fungi.

Host Nutrition

High nitrogen fertility (> 1/2 lb. N/1000 sq.ft./month) may increase disease severity. Low nitrogen may increase the frequency of occurrence of fairy ring.

Diagnosis

Circles or archs of mushrooms or wilted, dead or dark green turf. White mats of fungal mycelium may be found in thatch or soil associated with circles or archs.

Environment

Light to moderate textured soils.
Soil pH of 5 to 7.5.
Low to moderate soil moisture.

Fairy Ring

Disease Forecasting and Pathogen Detection

A fairy ring forecaster and detection kits are not available.

Chemical Controls

Soil fumigants* –
methyl bromide, chloropicrin, formaldehyde, Vorlex, Vapam.

Fungicides – Prostar.

Resistant Species and Cultivars

No turfgrasses are known to be resistant to fairy ring fungi.

Management

Cultural Controls

Maintain moderate nitrogen fertility (1/2 lb. N/1000 sq.ft./month).
Maintain moderate to high levels of phosphorus and potash according to soil tests.
Excavate ring and soil 12 inches deep and 24 inches beyond ring or arch and replace with new soil.
Remove sod, cultivate soil 6 to 8 inches deep in several directions, add wetting agent to soil, reseed or sod.

* These chemicals are highly toxic to turfgrasses, animals and other life forms.

Fusarium Patch

Hosts

All cool season turfgrasses. Annual bluegrass, bentgrasses and perennial ryegrass are particularly susceptible.

Pathogen

Microdochium nivale

Host Nutrition

Disease is severe under high nitrogen fertility (> 1/2 lb. N/1000 sq.ft./month) during late summer, fall, winter and spring. Disease may be severe on soils low in phosphorus and potash.

Diagnosis

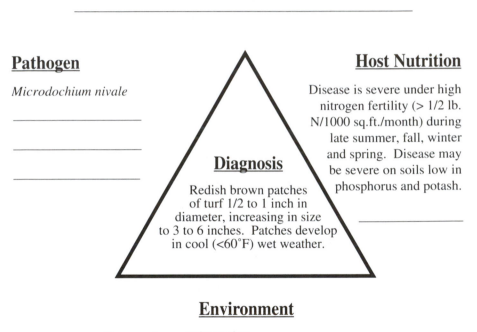

Redish brown patches of turf 1/2 to 1 inch in diameter, increasing in size to 3 to 6 inches. Patches develop in cool (<60˚F) wet weather.

Environment

Temperature <60˚F (16˚C).
More than 10 hrs. of leaf wetness per day for several days.
Disease severity may be increased by applications of lime.

Fusarium Patch

Disease Forecasting and Pathogen Detection

A Fusarium patch forecaster and detection kits are not available.

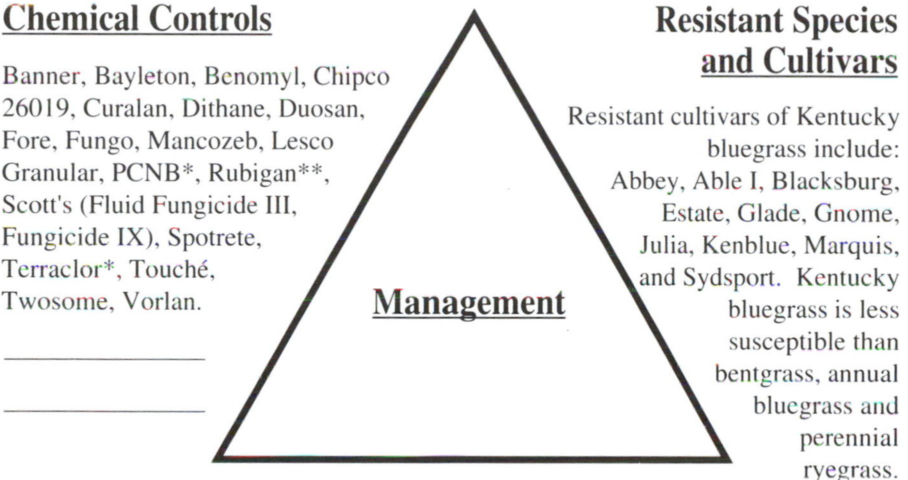

Chemical Controls

Banner, Bayleton, Benomyl, Chipco 26019, Curalan, Dithane, Duosan, Fore, Fungo, Mancozeb, Lesco Granular, PCNB*, Rubigan**, Scott's (Fluid Fungicide III, Fungicide IX), Spotrete, Terraclor*, Touché, Twosome, Vorlan.

____

____

Resistant Species and Cultivars

Resistant cultivars of Kentucky bluegrass include: Abbey, Able I, Blacksburg, Estate, Glade, Gnome, Julia, Kenblue, Marquis, and Sydsport. Kentucky bluegrass is less susceptible than bentgrass, annual bluegrass and perennial ryegrass.

Cultural Controls

Avoid high nitrogen (>1/2 lb. N/1000 sq.ft./month) in late summer and early fall.
Maintain moderate to high levels of phosphorous and potash according to soil tests.
Decrease shade and increase air circulation to enhance drying of turf.
Avoid applications of lime if possible.
Avoid irrigation in late afternoon and in evening prior to midnight.
Avoid mowing at height < 2 inches.

____

____

____

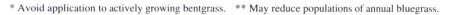

* Avoid application to actively growing bentgrass. ** May reduce populations of annual bluegrass.

Leaf Spot

Hosts

Bluegrasses, bentgrasses, fescues and perennial ryegrass.

Pathogen

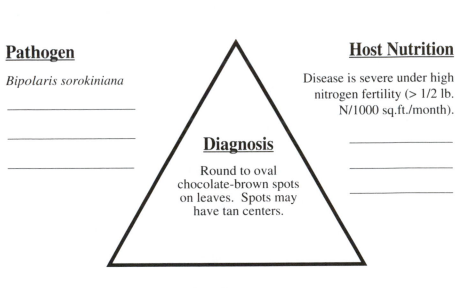

Bipolaris sorokiniana

Host Nutrition

Disease is severe under high nitrogen fertility (> 1/2 lb. N/1000 sq.ft./month).

Diagnosis

Round to oval chocolate-brown spots on leaves. Spots may have tan centers.

Environment

Temperatures of 77°- 95°F (25°- 35°C).
Disease severity increases with increases in temperature.
More than 10 hrs. of leaf wetness per day for several days.

Leaf Spot

Disease Forecasting and Pathogen Detection

A leaf spot forecaster and detection kits are not available.

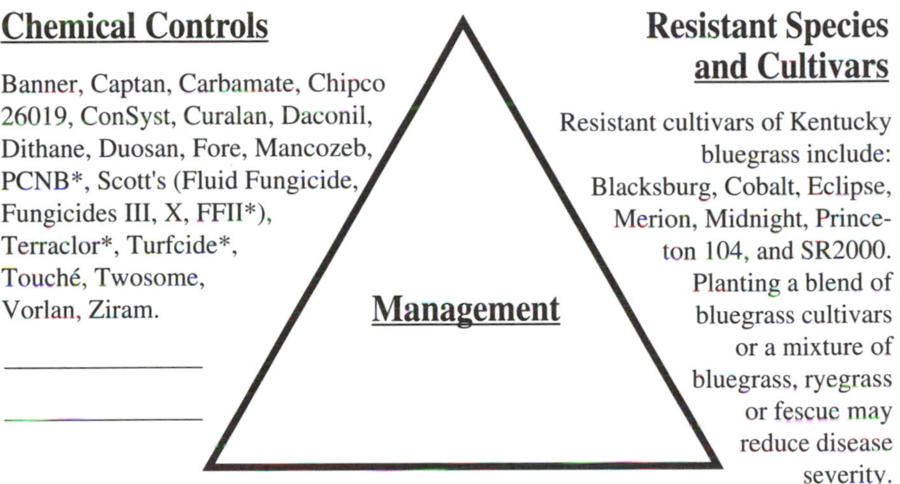

Chemical Controls

Banner, Captan, Carbamate, Chipco 26019, ConSyst, Curalan, Daconil, Dithane, Duosan, Fore, Mancozeb, PCNB*, Scott's (Fluid Fungicide, Fungicides III, X, FFII*), Terraclor*, Turfcide*, Touché, Twosome, Vorlan, Ziram.

____

____

Management

Resistant Species and Cultivars

Resistant cultivars of Kentucky bluegrass include: Blacksburg, Cobalt, Eclipse, Merion, Midnight, Princeton 104, and SR2000. Planting a blend of bluegrass cultivars or a mixture of bluegrass, ryegrass or fescue may reduce disease severity.

Cultural Controls

Apply moderate amounts of nitrogen during summer (1/4-1/2 lb. N/ 1000 sq.ft./month).
Maintain moderate to high levels of soil P and K.
Decrease shade and increase air circulation to enhance drying of turf.
Avoid irrigaiton in late afternoon and in evening prior to midnight.
Limit thatch to 1/2 inch or less.
Raise mowing height.
Use light-weight mowing equipment to reduce stress.
Avoid mowing at height < 2 inches.

____

____

____

* Avoid application to bentgrasses.

Melting-Out

Hosts

Bluegrasses, ryegrasses and tall fescues.

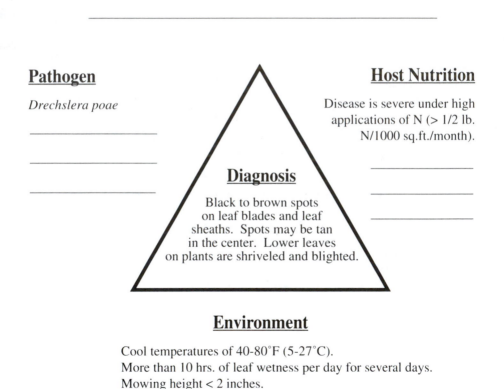

Pathogen

Drechslera poae

Host Nutrition

Disease is severe under high applications of N (> 1/2 lb. N/1000 sq.ft./month).

Diagnosis

Black to brown spots on leaf blades and leaf sheaths. Spots may be tan in the center. Lower leaves on plants are shriveled and blighted.

Environment

Cool temperatures of 40-80°F (5-27°C).
More than 10 hrs. of leaf wetness per day for several days.
Mowing height < 2 inches.

Melting-Out

Disease Forecasting and Pathogen Detection

A melting-out forecaster and detection kits are not available.

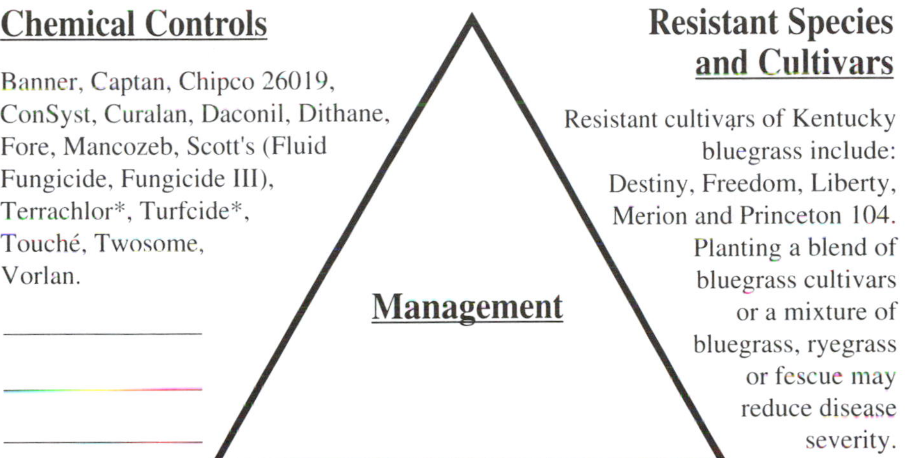

Chemical Controls

Banner, Captan, Chipco 26019, ConSyst, Curalan, Daconil, Dithane, Fore, Mancozeb, Scott's (Fluid Fungicide, Fungicide III), Terrachlor*, Turfcide*, Touché, Twosome, Vorlan.

Resistant Species and Cultivars

Resistant cultivars of Kentucky bluegrass include: Destiny, Freedom, Liberty, Merion and Princeton 104. Planting a blend of bluegrass cultivars or a mixture of bluegrass, ryegrass or fescue may reduce disease severity.

Cultural Controls

Fertilize with low to moderate levels of N during spring, summer and fall (1/4-1/2 lb. N/1000 sq.ft./month).
Decrease shade and increase air circulation to enhance drying of turf.
Avoid irrigation in late afternoon and in evening prior to midnight.
Avoid mowing at height < 2 inches.
Use light-weight mowing equipment to avoid stress on turf.
Limit thatch thickness to 1/2 inch or less.

* Avoid applications to bentgrasses.

Necrotic Ring Spot

Hosts

Kentucky bluegrass and fine-leaf fescues.

Pathogen

Leptosphaeria korrae

Host Nutrition

Disease may be severe when high nitrogen (> 1/2 lb. N/1000 sq.ft./ month) is applied, especially in spring and summer.

Diagnosis

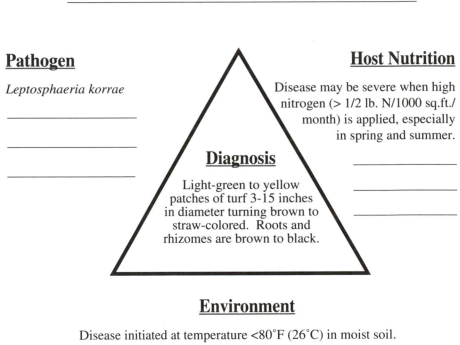

Light-green to yellow patches of turf 3-15 inches in diameter turning brown to straw-colored. Roots and rhizomes are brown to black.

Environment

Disease initiated at temperature <80°F (26°C) in moist soil.
Severity of symptoms increases with drought and high temperatures (>80F).
Disease is severe on compacted soils.

Necrotic Ring Spot

Disease Forecasting and Pathogen Detection

A necrotic ring spot forecaster and detection kits are not available.

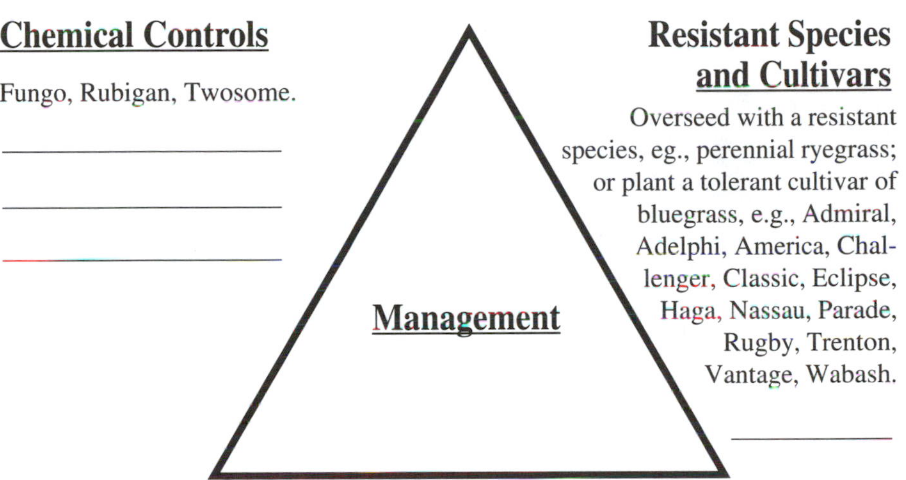

Chemical Controls

Fungo, Rubigan, Twosome.

Resistant Species and Cultivars

Overseed with a resistant species, eg., perennial ryegrass; or plant a tolerant cultivar of bluegrass, e.g., Admiral, Adelphi, America, Challenger, Classic, Eclipse, Haga, Nassau, Parade, Rugby, Trenton, Vantage, Wabash.

Cultural Controls

Avoid high amounts of "fast-release" nitrogen.
Maintain moderate to high levels of phosphorus and potash according to soil tests.
Avoid drought stress.
Raise mowing height to at least 2 inches.
Avoid soil compaction – top-dress and aerify as needed, use lightweight equipment.
Reduce thatch thickness to 1/2 inch or less.

Nematodes

Hosts

All common species of turfgrasses.

Pathogen

More than 25 different species attack cool season turfgrasses.

Host Nutrition

Symptoms are often severe on lush, excessively fertilized turf or on thin, under-fertilized turf.

Diagnosis

Irregular shaped light green or yellow patches of turf up to several feet in diameter. Leaves may be yellow or brown from the tip. Roots may be thin, stunted or knotted.

Environment

Soil temperatures > 40˚F (5˚C).
Symptoms are often severe on turf growing in sandy, light-textured soils.
Symptoms may be enhanced by drought and high temperatures (>80˚F, 26˚C).

Nematode

Disease Forecasting and Pathogen Detection

A nematode forecaster and detection kits are not available.

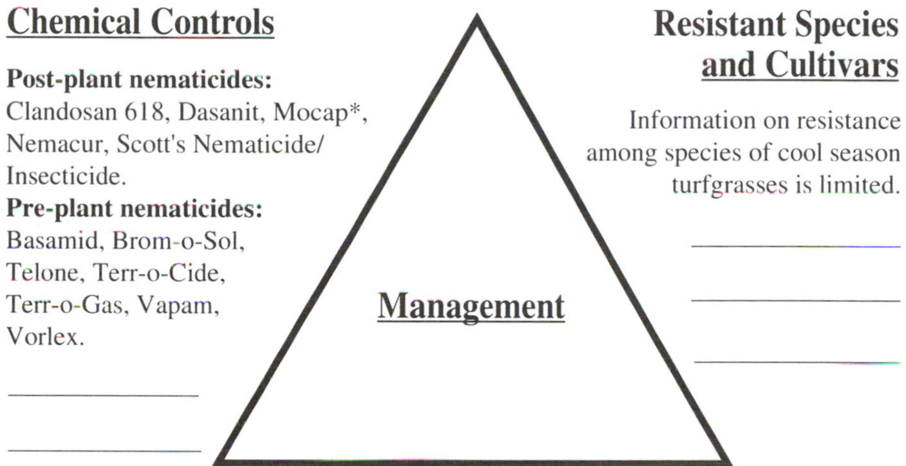

Chemical Controls

Post-plant nematicides:
Clandosan 618, Dasanit, Mocap*, Nemacur, Scott's Nematicide/ Insecticide.
Pre-plant nematicides:
Basamid, Brom-o-Sol, Telone, Terr-o-Cide, Terr-o-Gas, Vapam, Vorlex.

Resistant Species and Cultivars

Information on resistance among species of cool season turfgrasses is limited.

Cultural Controls

Maintain a balanced fertility program.
Apply 1/2 lb. N/1000 sq.ft./month during spring and summer.
Maintain moderate to high levels of phosphorus and potash according to soil tests.
Have soil analyzed for nematodes prior to seeding or sodding.
Use sod that is nematode-free.

* Avoid applications to bentgrasses.

Nigrospora Blight

Hosts

Kentucky bluegrass, fescues and perennial ryegrass.

Pathogen

Nigrospora sphaerica

Host Nutrition

Disease may be severe on low nutrient turf.

Diagnosis

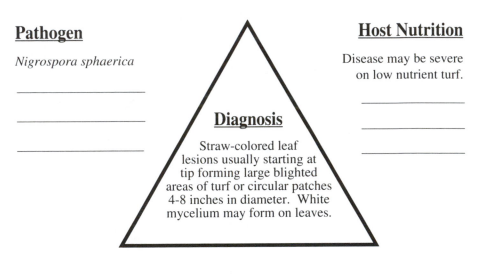

Straw-colored leaf lesions usually starting at tip forming large blighted areas of turf or circular patches 4-8 inches in diameter. White mycelium may form on leaves.

Environment

Temperature > 80°F (27°C).
More than 10 hrs. of leaf wetness per day for several days.
Disease may be severe on turf subjected to drought, herbicide stress or low mowing height.

Nigrospora Blight

Disease Forecasting and Pathogen Detection

A Nigrospora blight forecaster and detection kits are not available.

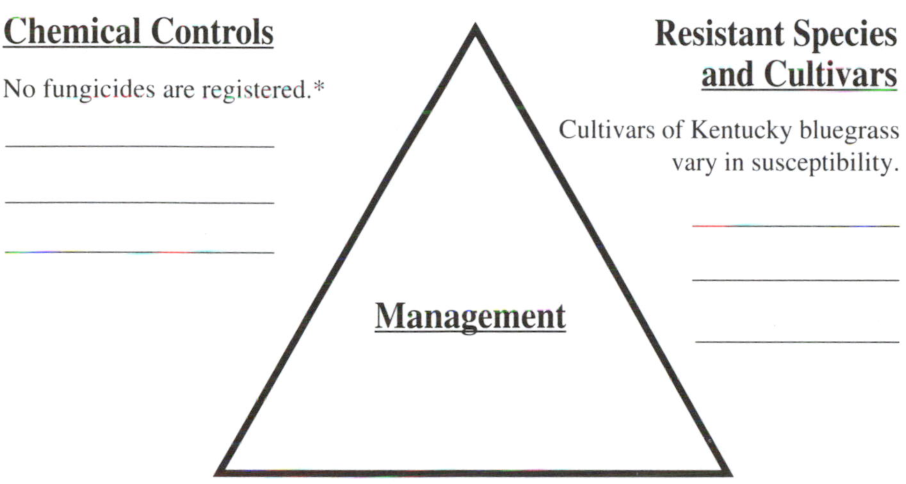

Chemical Controls

No fungicides are registered.*

Resistant Species and Cultivars

Cultivars of Kentucky bluegrass vary in susceptibility.

Management

Cultural Controls

Maintain moderate nitrogen fertility (1/2 lb. N/1000 sq.ft./month) during summer.
Maintain moderate to high levels of phosphorus and potash according to soil tests.
Avoid drought stress.
Avoid irrigation in late afternoon and evening prior to midnight.
Decrease shade and increase air circulation to enhance drying of turf.
Maintain turf at height of 2 inches or greater.
Avoid herbicide application during summer.

* Chipco 26019 and Daconil 2787 provided control in experimental evaluations.

Pink Patch

Hosts

Perennial ryegrass, fine-leaf fescues, bentgrasses and bluegrasses.

Pathogen

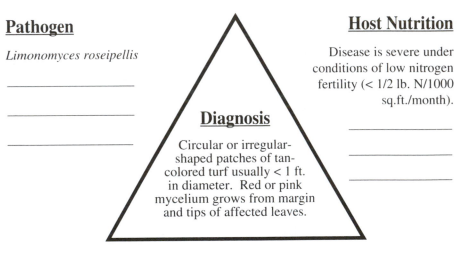

Limonomyces roseipellis

Host Nutrition

Disease is severe under conditions of low nitrogen fertility (< 1/2 lb. N/1000 sq.ft./month).

Diagnosis

Circular or irregular-shaped patches of tan-colored turf usually < 1 ft. in diameter. Red or pink mycelium grows from margin and tips of affected leaves.

Environment

Cool temperatures of 60°-75°F (16°-24°C).
More than 10 hrs. of leaf wetness per day for several days.

Pink Patch

Disease Forecasting and Pathogen Detection

A pink patch forecaster and detection kits are not available.

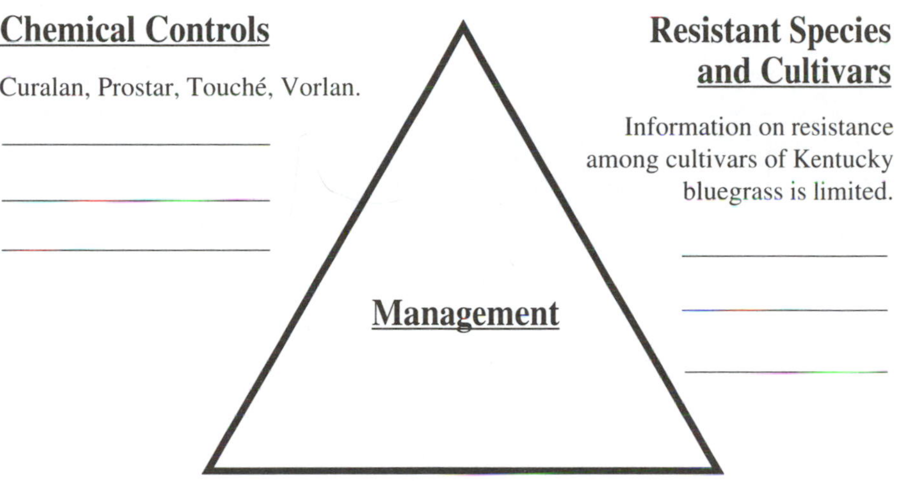

Chemical Controls

Curalan, Prostar, Touché, Vorlan.

Resistant Species and Cultivars

Information on resistance among cultivars of Kentucky bluegrass is limited.

Management

Cultural Controls

Avoid low fertility.
Apply at least 1/2 lb. N/1000 sq.ft./month.
Maintain moderate to high levels of phosphorus and potash according to soil tests.
Reduce shade and increase air circulation to enhance drying of turf.
Avoid irrigation in the late afternoon or in evening prior to midnight.
Mow turf at least once per week to remove diseased portions of leaf blades.

Pink Snow Mold

Hosts

All cool season grasses. Bentgrasses, annual bluegrass and perennial ryegrass are particularly susceptible.

Pathogen

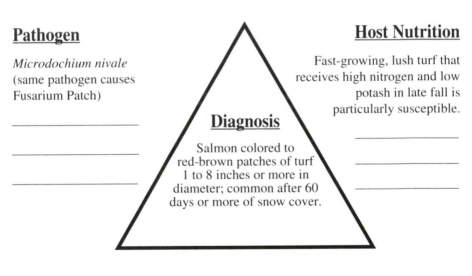

Microdochium nivale (same pathogen causes Fusarium Patch)

Host Nutrition

Fast-growing, lush turf that receives high nitrogen and low potash in late fall is particularly susceptible.

Diagnosis

Salmon colored to red-brown patches of turf 1 to 8 inches or more in diameter; common after 60 days or more of snow cover.

Environment

Disease is common after at least 60 days of snow cover, but pathogen can infect turf in absence of snow (see Fusarium Patch). Disease is particularly severe when snow covers unfrozen ground.

Pink Snow Mold

Disease Forecasting and Pathogen Detection

A pink snow mold forecaster and detection kits are not available.

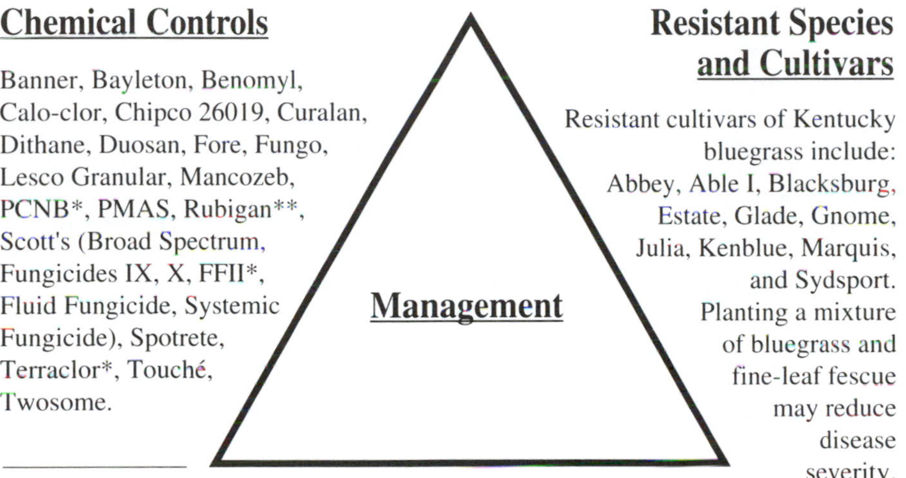

Chemical Controls

Banner, Bayleton, Benomyl, Calo-clor, Chipco 26019, Curalan, Dithane, Duosan, Fore, Fungo, Lesco Granular, Mancozeb, PCNB*, PMAS, Rubigan**, Scott's (Broad Spectrum, Fungicides IX, X, FFII*, Fluid Fungicide, Systemic Fungicide), Spotrete, Terraclor*, Touché, Twosome.

Management

Resistant Species and Cultivars

Resistant cultivars of Kentucky bluegrass include: Abbey, Able I, Blacksburg, Estate, Glade, Gnome, Julia, Kenblue, Marquis, and Sydsport. Planting a mixture of bluegrass and fine-leaf fescue may reduce disease severity.

Cultural Controls

Maintain moderate nitrogen fertility (1/2 lb. N/1000 sq.ft./month) during late summer and fall.
Maintain high potash levels according to soil tests.
Use snow fence, shrubs or knolls as wind-breaks to prevent excess snow from accumulating.
Prevent snow compaction by machinery or skiers.
Melt snow in spring with organic fertilizers.
Physically remove snow in spring.
Follow controls for Fusarium Patch after snow melt.

* Avoid application to actively growing bentgrasses. ** May reduce populations of annual bluegrass.

Powdery Mildew

Hosts

Kentucky bluegrass and fine-leaf fescues.

Pathogen

Erysiphe graminis

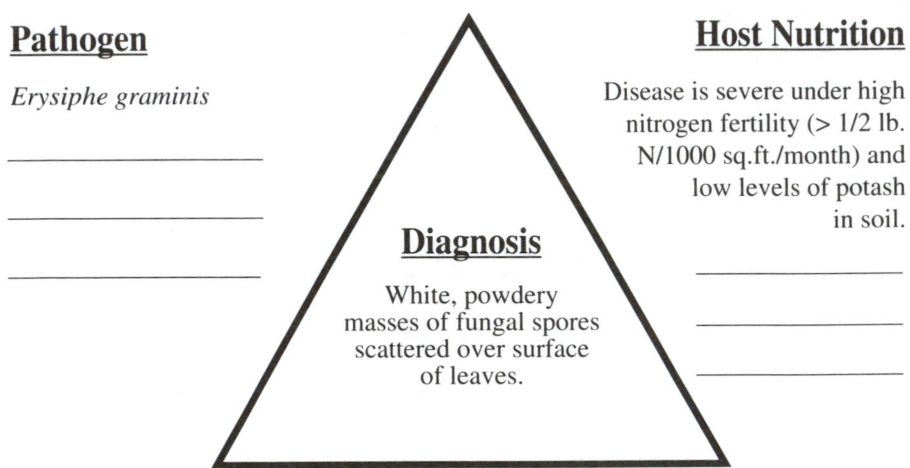

Diagnosis

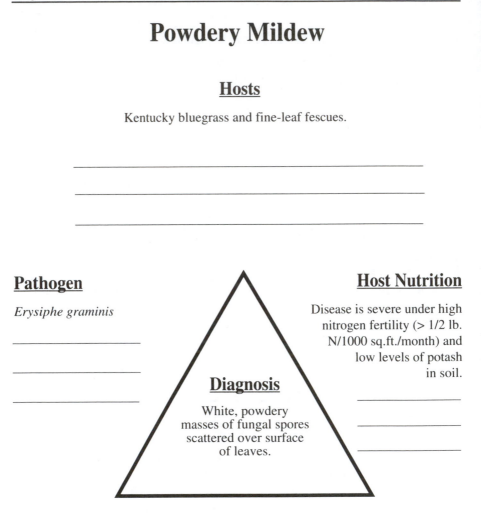

White, powdery masses of fungal spores scattered over surface of leaves.

Host Nutrition

Disease is severe under high nitrogen fertility (> 1/2 lb. N/1000 sq.ft./month) and low levels of potash in soil.

Environment

Disease is severe in shaded areas at temperatures of 60°-72°F (15°-22°C). High humidity is required for infection, but leaf wetness is not essential.

Powdery Mildew

Disease Forecasting and Pathogen Detection

A powdery mildew forecaster and detection kits are not available.

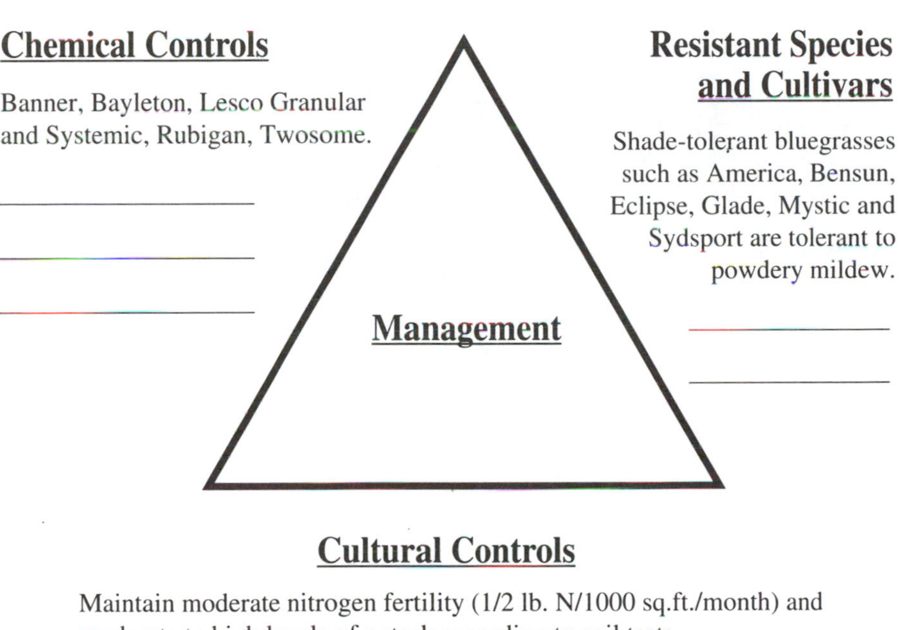

Chemical Controls

Banner, Bayleton, Lesco Granular and Systemic, Rubigan, Twosome.

Resistant Species and Cultivars

Shade-tolerant bluegrasses such as America, Bensun, Eclipse, Glade, Mystic and Sydsport are tolerant to powdery mildew.

Cultural Controls

Maintain moderate nitrogen fertility (1/2 lb. N/1000 sq.ft./month) and moderate to high levels of potash according to soil tests.
Reduce shade and increase air circulation.

Pythium Blight

Hosts

All cool season grasses. Annual bluegrass and perennial ryegrass are particularly susceptible.

Pathogen

Pythium aphanidermatum and other species of *Pythium.*

Host Nutrition

Disease is severe under high nitrogen fertility (> 1/2 lb. N/1000 sq.ft./month). Deficiency in calcium may increase susceptibility.

Diagnosis

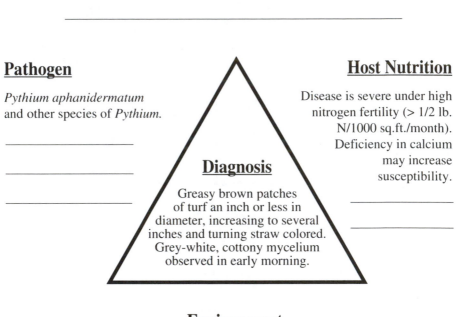

Greasy brown patches of turf an inch or less in diameter, increasing to several inches and turning straw colored. Grey-white, cottony mycelium observed in early morning.

Environment

Night temperature > 65°F (18°C).
More than 10 hrs. of leaf wetness per day for several days.
Poor surface and sub-surface drainage.

Pythium Blight

Disease Forecasting and Pathogen Detection

Pythium blight forecasters are available from Pest Management Supply, P.O. Box 936, Amherst, MA 01004, or Neogen Corp., 620 Lesher Pl., Lansing, MI 48912. Detection kits are available from Neogen Corp.

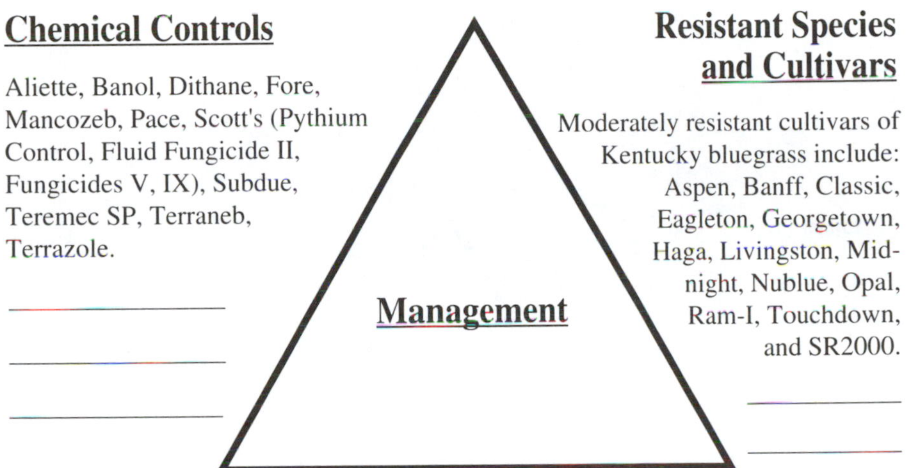

Chemical Controls

Aliette, Banol, Dithane, Fore, Mancozeb, Pace, Scott's (Pythium Control, Fluid Fungicide II, Fungicides V, IX), Subdue, Teremec SP, Terraneb, Terrazole.

Resistant Species and Cultivars

Moderately resistant cultivars of Kentucky bluegrass include: Aspen, Banff, Classic, Eagleton, Georgetown, Haga, Livingston, Midnight, Nublue, Opal, Ram-I, Touchdown, and SR2000.

Cultural Controls

Maintain moderate nitrogen fertility (1/2 lb. N/1000 sq.ft./month).
Maintain optimum plant calcium levels.
Decrease shade and increase air circulation to enhance drying of turf.
Improve surface and subsurface drainage.
Avoid mowing susceptible areas when turf is wet, particularly when night temperatures are > 70˚F (21˚C).
Avoid mowing at height < 2 inches.

Pythium Root Rot

Hosts

All species of cool season grasses.

Pathogen

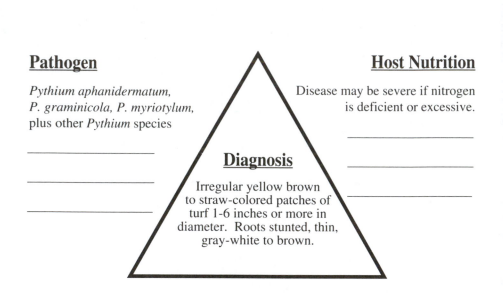

Pythium aphanidermatum, P. graminicola, P. myriotylum, plus other *Pythium* species

Host Nutrition

Disease may be severe if nitrogen is deficient or excessive.

Diagnosis

Irregular yellow brown to straw-colored patches of turf 1-6 inches or more in diameter. Roots stunted, thin, gray-white to brown.

Environment

Cool (32°-50°F, 0-10°C) or warm (70°-90°F, 21°-32°C) soil temperatures*.
High soil moisture.
Poor surface or subsurface drainage.
Conditions unfavorable for carbohydrate development by leaves – low light, low mowing height, excessive wear.

* Some *Pythium* species are favored by cool soils, other species by warm soils.

Pythium Root Rot

Disease Forecasting and Pathogen Detection

A Pythium root rot forecaster is not available. Detection kits are available from Neogen Corp., 620 Lesher Pl., Lansing, MI 48912.

Chemical Controls

Banol, Koban, Subdue, Truban.

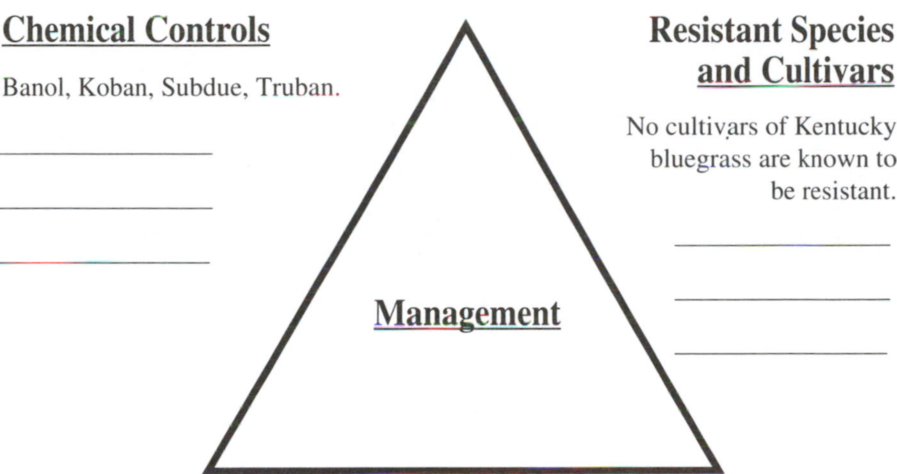

Resistant Species and Cultivars

No cultivars of Kentucky bluegrass are known to be resistant.

Cultural Controls

Maintain moderate levels of nitrogen (1/2 lb. N/1000 sq.ft./month).
Do not over fertilize with nitrogen in spring when roots are forming.
Maintain moderate to high levels of phosphorus and potash according to soil tests.
Improve surface and subsurface drainage.
Raise mowing height to at least 2 inches.
Decrease shade.
Use light-weight mowing equipment.
Applications of certain composts may reduce disease severity.

Red Thread

Hosts

Bluegrasses, bentgrasses, fine-leaf fescues and perennial ryegrass.
Fine-leaf fescues and perennial ryegrass are particularly susceptible.

Pathogen

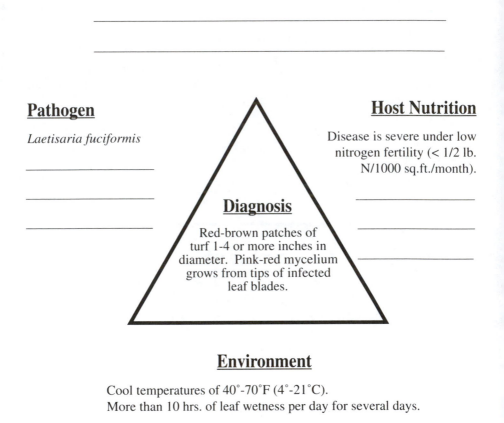

Laetisaria fuciformis

Host Nutrition

Disease is severe under low nitrogen fertility (< 1/2 lb. N/1000 sq.ft./month).

Diagnosis

Red-brown patches of turf 1-4 or more inches in diameter. Pink-red mycelium grows from tips of infected leaf blades.

Environment

Cool temperatures of 40°-70°F (4°-21°C).
More than 10 hrs. of leaf wetness per day for several days.

Red Thread

Disease Forecasting and Pathogen Detection

A red thread forecaster and detection kits are not available.

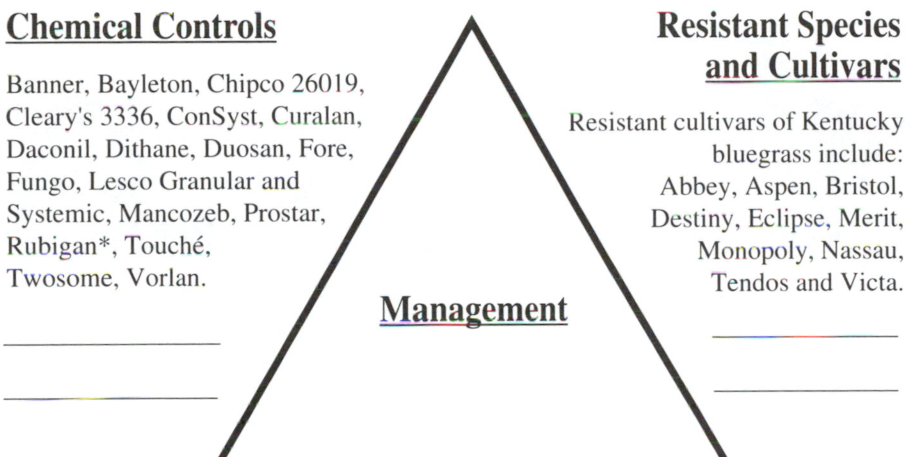

Chemical Controls

Banner, Bayleton, Chipco 26019, Cleary's 3336, ConSyst, Curalan, Daconil, Dithane, Duosan, Fore, Fungo, Lesco Granular and Systemic, Mancozeb, Prostar, Rubigan*, Touché, Twosome, Vorlan.

Resistant Species and Cultivars

Resistant cultivars of Kentucky bluegrass include: Abbey, Aspen, Bristol, Destiny, Eclipse, Merit, Monopoly, Nassau, Tendos and Victa.

Cultural Controls

Fertilize with at least 1/2 lb. N/1000 sq.ft./month.
Maintain moderate to high levels of potash and phosphorus according to soil tests.
Reduce shade and increase air circulation to enhance drying of turf.
Avoid irrigation in late afternoon or in evening prior to midnight.
Maintain soil pH at 6.5 to 7.0.
Mow turf at least once per week to remove diseased portions of leaf blades.

* May reduce populations of annual bluegrass.

Rhizoctonia Leaf and Sheath Spot

Hosts

Bentgrasses, bluegrasses, perennial ryegrass and tall fescue.

Pathogen

Rhizoctonia zeae

Host Nutrition

Effects of nutrition on disease development are unknown.

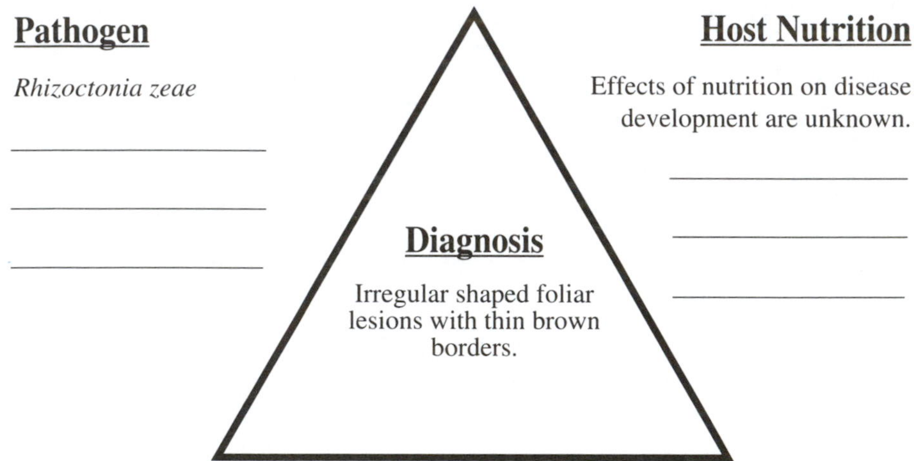

Diagnosis

Irregular shaped foliar lesions with thin brown borders.

Environment

Day temperatures > 90˚F (32˚C).
More than 10 hrs. of leaf wetness per day for several days.

Rhizoctonia Leaf and Sheath Spot

Disease Forecasting and Pathogen Detection

A leaf and sheath spot forecaster and detection kits are not available.

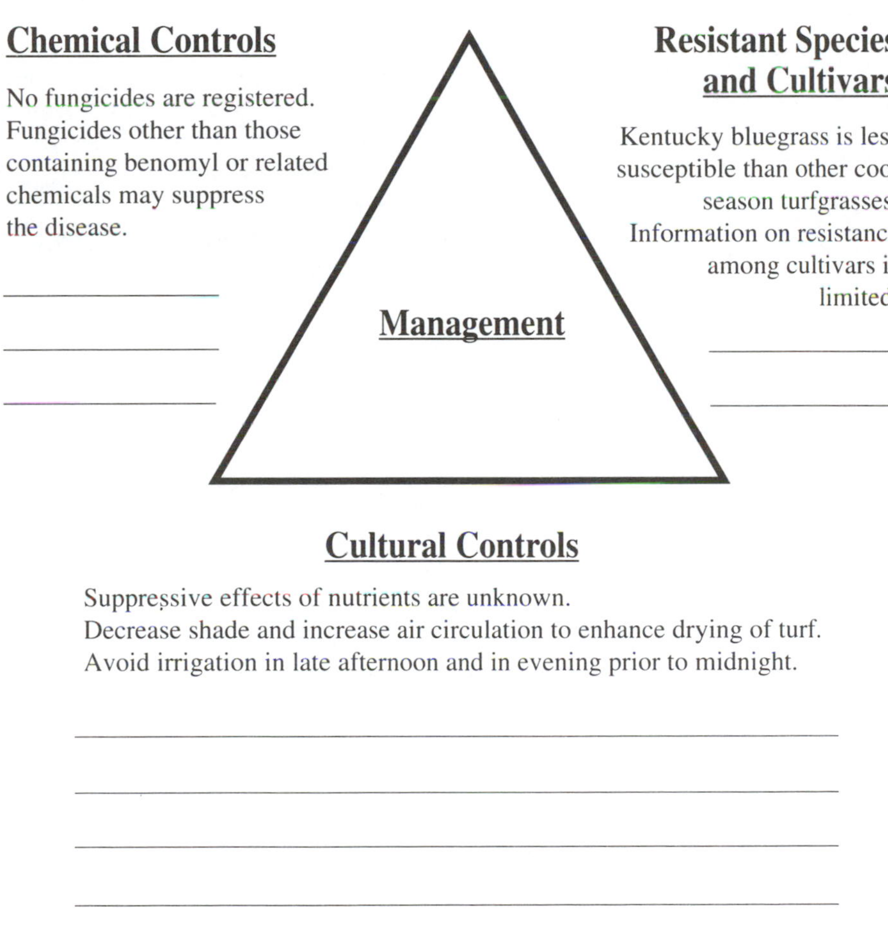

Chemical Controls

No fungicides are registered. Fungicides other than those containing benomyl or related chemicals may suppress the disease.

Resistant Species and Cultivars

Kentucky bluegrass is less susceptible than other cool season turfgrasses. Information on resistance among cultivars is limited.

Cultural Controls

Suppressive effects of nutrients are unknown.
Decrease shade and increase air circulation to enhance drying of turf.
Avoid irrigation in late afternoon and in evening prior to midnight.

Rust

Hosts

All common species of cool season turfgrasses.

Pathogen

Several species of *Puccinia* and *Uromyces*

Host Nutrition

Disease is severe on turf grown under low fertility.

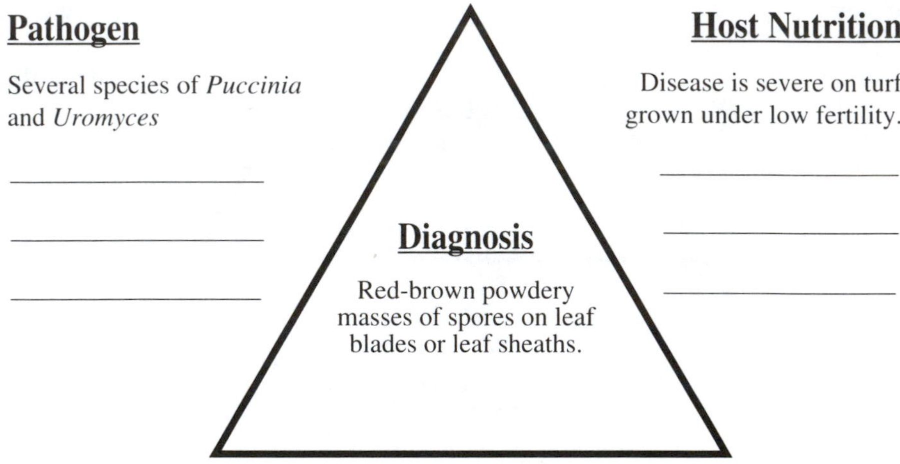

Diagnosis

Red-brown powdery masses of spores on leaf blades or leaf sheaths.

Environment

Temperatures of 68°-86°F (20°-30°C).
Disease is severe on turf subjected to drought stress, low mowing, shade or poor air circulation.

Rust

Disease Forecasting and Pathogen Detection

A rust forecaster and detection kits are not available.

Chemical Controls

Banner, Bayleton, Captan, Carbamate, Cleary's 3336, ConSyst, Daconil, Dithane, Duosan, Fore, Mancozeb, Rubigan*, Scott's (Fluid Fungicide III, Fungicide VII, FFII**), Twosome, Ziram.

Management

Resistant Species and Cultivars

Cultivars of Kentucky bluegrass that are resistant to one or more rust fungi include: Alpine, Barzan, Bristol, Dawn, Destiny, Georgetown, Haga, Nassau, Suffolk, Sydsport, Tendos and Trenton.

Cultural Controls

Maintain moderate and balanced fertility throughout the growing season.
Reduce shade and increase air circulation to enhance drying of turf.
Increase mowing height to at least 2 inches.
Avoid drought stress.
Avoid irrigation in late afternoon and in evening prior to midnight.

* May reduce populations of annual bluegrass. ** Avoid application to bentgrasses.

Stripe Smut

Hosts

Bluegrasses, bentgrasses, perennial ryegrass and tall fescue. Kentucky bluegrass is more susceptible than other cool season grasses.

Pathogen

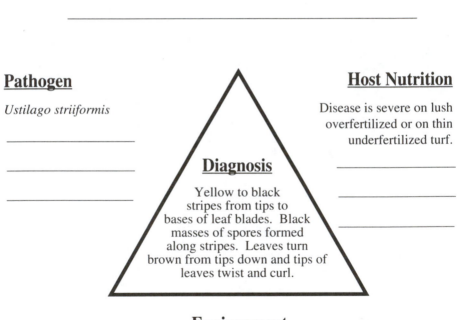

Ustilago striiformis

Host Nutrition

Disease is severe on lush overfertilized or on thin underfertilized turf.

Diagnosis

Yellow to black stripes from tips to bases of leaf blades. Black masses of spores formed along stripes. Leaves turn brown from tips down and tips of leaves twist and curl.

Environment

Infection occurs at 50°- 68°F (10°-20°C).
Severe symptoms evident during drought and temperatures >75°F (24°C).
Symptoms often more severe on acid soils and on turf with excessive thatch (> 1/2 inch thick).

Stripe Smut

Disease Forecasting and Pathogen Detection

A stripe smut forecaster and detection kits are not available.

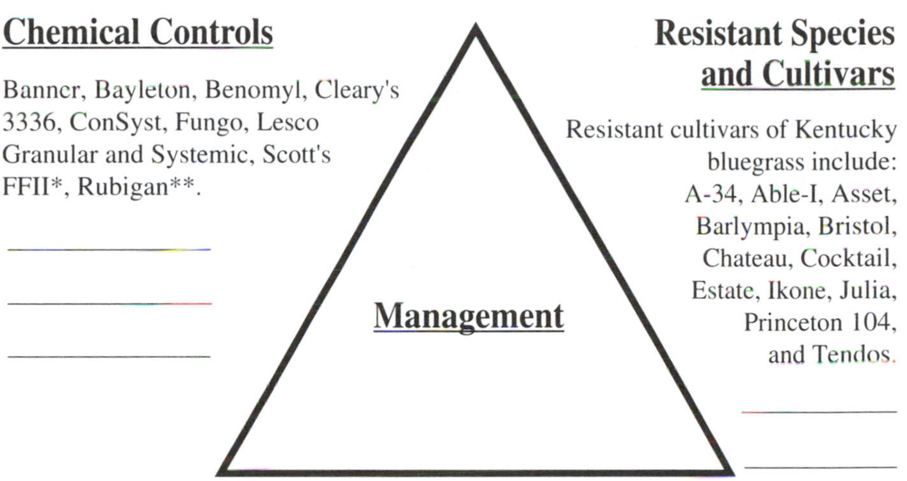

Chemical Controls

Banner, Bayleton, Benomyl, Cleary's 3336, ConSyst, Fungo, Lesco Granular and Systemic, Scott's FFII*, Rubigan**.

Resistant Species and Cultivars

Resistant cultivars of Kentucky bluegrass include: A-34, Able-I, Asset, Barlympia, Bristol, Chateau, Cocktail, Estate, Ikone, Julia, Princeton 104, and Tendos.

Cultural Controls

Maintain moderate nitrogen fertility (1/2 lb. N/1000 sq.ft./month).
Maintain moderate phosphorus and high potash levels according to soil tests.
Avoid drought stress.
Apply lime if soil pH < 6.
Dethatch turf if thatch is > 1/2 inch thick.
Avoid mowing at height < 2 inches.

* Avoid applications to bentgrasses. ** May reduce populations of annual bluegrass.

Summer Patch

Hosts

Bluegrasses and fine-leaf fescues.

Pathogen

Magnaporthe poae

Host Nutrition

Disease may be severe when turf is fertilized with fast-release sources of nitrogen.

Diagnosis

Circular patches of wilted to straw-colored turf, usually less than 10 inches in diameter. Leaves turn yellow or brown starting at tips. Roots are light to dark brown.

Environment

Day-time temperature > 85˚F (29˚C).
High soil moisture.
Poor surface or subsurface drainage.
Low mowing height.

Summer Patch

Disease Forecasting and Pathogen Detection

A summer patch forecaster and detection kits are not available.

Chemical Controls

Banner, Bayleton, Fungo, Lesco Granular and Systemic, Rubigan*, Twosome.

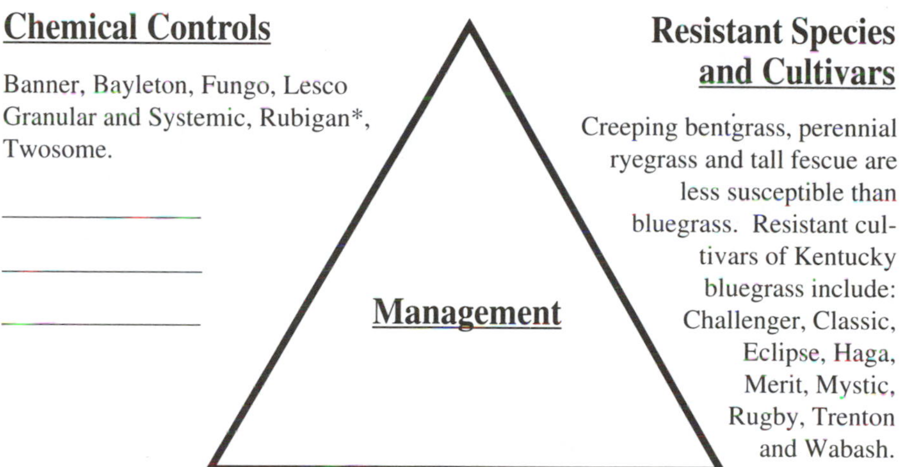

Resistant Species and Cultivars

Creeping bentgrass, perennial ryegrass and tall fescue are less susceptible than bluegrass. Resistant cultivars of Kentucky bluegrass include: Challenger, Classic, Eclipse, Haga, Merit, Mystic, Rugby, Trenton and Wabash.

Cultural Controls

Avoid "fast-release" sources of nitrogen.
Do not apply > 1/2 lb. N/1000 sq.ft./month during spring and summer.
Improve surface and subsurface drainage.
Reduce compaction.
Syringe turf with water when temperature > 85˚F (29˚C).
Raise mowing height to at least 2 inches.
Use light-weight mowing equipment.

* May reduce populations of annual bluegrass.

Typhula Blight
(Gray Snow Mold)

Hosts

All cool season turfgrasses. Bentgrasses, annual bluegrass and perennial ryegrass are particularly susceptible.

Pathogen

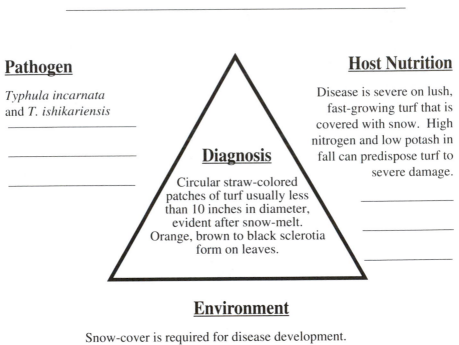

Typhula incarnata and *T. ishikariensis*

Host Nutrition

Disease is severe on lush, fast-growing turf that is covered with snow. High nitrogen and low potash in fall can predispose turf to severe damage.

Diagnosis

Circular straw-colored patches of turf usually less than 10 inches in diameter, evident after snow-melt. Orange, brown to black sclerotia form on leaves.

Environment

Snow-cover is required for disease development. Disease is severe when snow-cover exceeds 90 days.

Typhula Blight

Disease Forecasting and Pathogen Detection

A Typhula blight forecaster and detection kits are not available.

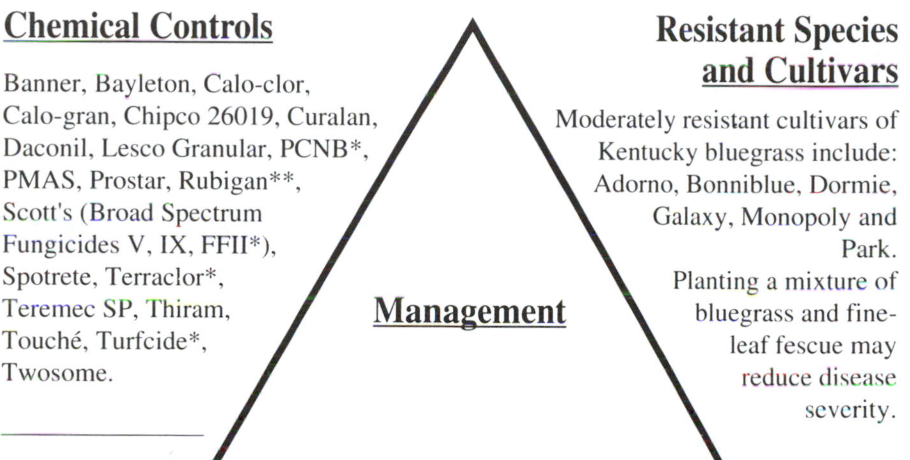

Chemical Controls

Banner, Bayleton, Calo-clor, Calo-gran, Chipco 26019, Curalan, Daconil, Lesco Granular, PCNB*, PMAS, Prostar, Rubigan**, Scott's (Broad Spectrum Fungicides V, IX, FFII*), Spotrete, Terraclor*, Teremec SP, Thiram, Touché, Turfcide*, Twosome.

Management

Resistant Species and Cultivars

Moderately resistant cultivars of Kentucky bluegrass include: Adorno, Bonniblue, Dormie, Galaxy, Monopoly and Park.

Planting a mixture of bluegrass and fine-leaf fescue may reduce disease severity.

Cultural Controls

Avoid a fertility program that results in lush, fast-growing turf in late fall and winter.

Maintain high potash levels according to soil test.

Use snow fence, hedges or knolls to prevent snow from accumulating excessively on turf.

Use dark-colored organic fertilizers or composts to melt snow in spring.

Physically remove snow in spring.

Prevent compaction of snow during winter.

* Avoid application to actively growing bentgrasses. ** May reduce populations of annual bluegrass.

White Patch
(White Blight)

Hosts

Fescues, bluegrasses and creeping bentgrass.

Pathogen

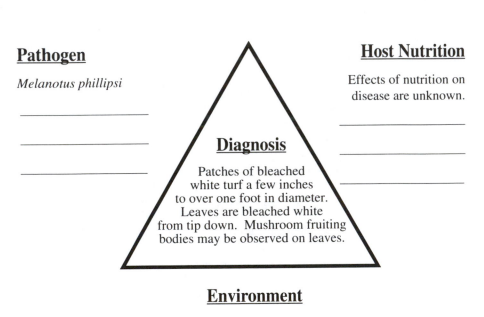

Melanotus phillipsi

Host Nutrition

Effects of nutrition on disease are unknown.

Diagnosis

Patches of bleached white turf a few inches to over one foot in diameter. Leaves are bleached white from tip down. Mushroom fruiting bodies may be observed on leaves.

Environment

Night temperatures > 70°F (21°C).
More than 10 hrs. of leaf wetness per day for several days.
Disease is particularly severe on soils from recently cleared forests.

White Patch

Disease Forecasting and Pathogen Detection

A white patch forecaster and detection kits are not available.

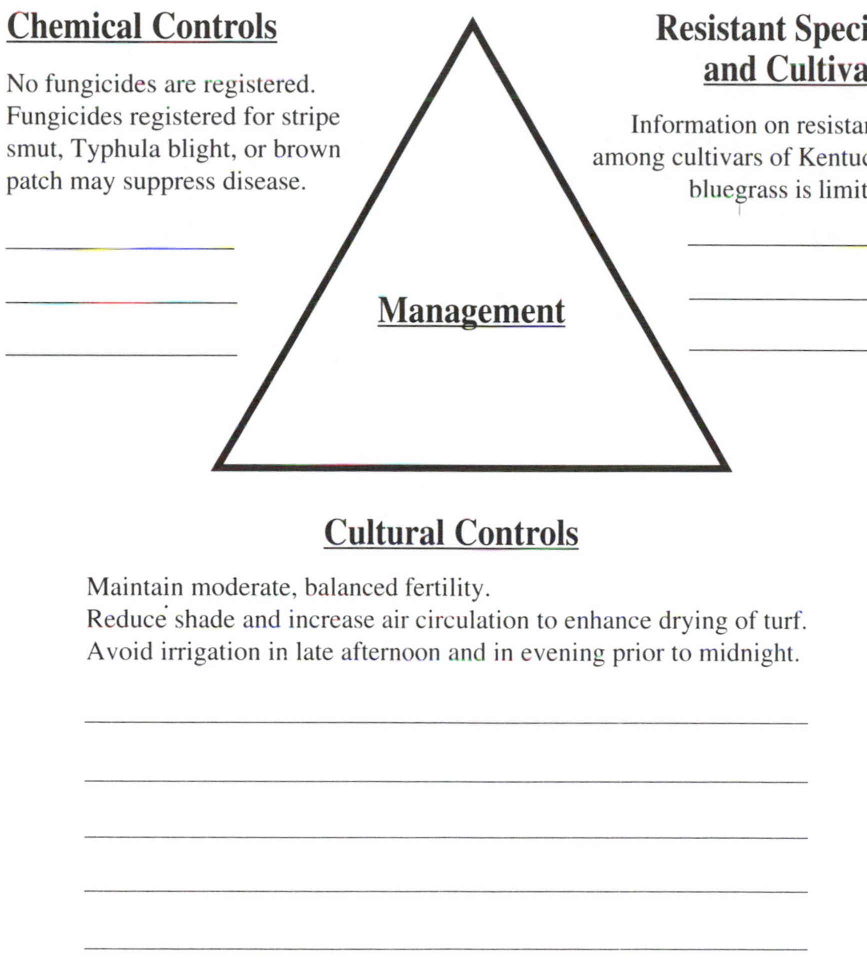

Chemical Controls

No fungicides are registered. Fungicides registered for stripe smut, Typhula blight, or brown patch may suppress disease.

Resistant Species and Cultivars

Information on resistance among cultivars of Kentucky bluegrass is limited.

Cultural Controls

Maintain moderate, balanced fertility.
Reduce shade and increase air circulation to enhance drying of turf.
Avoid irrigation in late afternoon and in evening prior to midnight.

Yellow Patch

Hosts

Bluegrasses, bentgrasses and tall fescue.

Pathogen

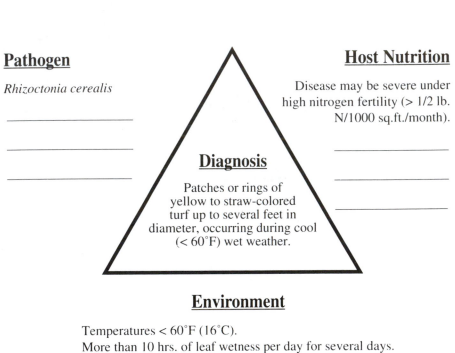

Rhizoctonia cerealis

Host Nutrition

Disease may be severe under high nitrogen fertility (> 1/2 lb. N/1000 sq.ft./month).

Diagnosis

Patches or rings of yellow to straw-colored turf up to several feet in diameter, occurring during cool (< 60°F) wet weather.

Environment

Temperatures < 60°F (16°C).
More than 10 hrs. of leaf wetness per day for several days.
Disease is severe on turf with excessive thatch.

Yellow Patch

Disease Forecasting and Pathogen Detection

A yellow patch forecaster and detection kits are not available.

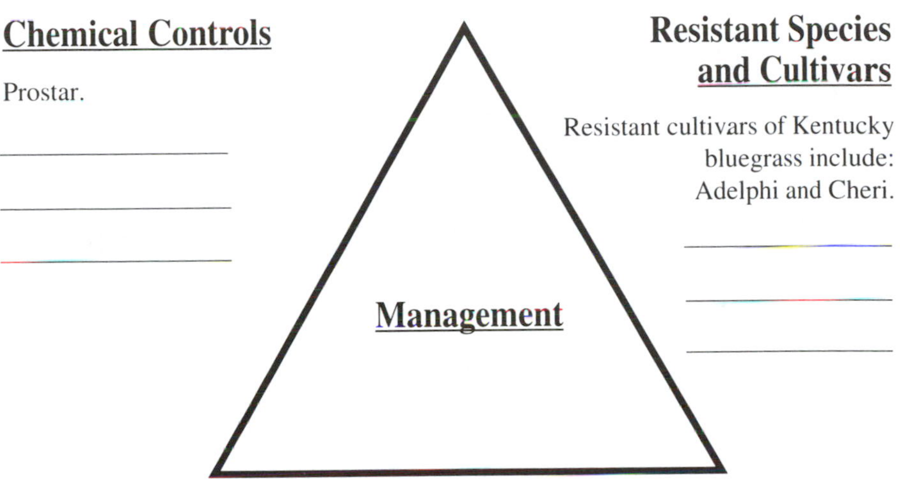

Chemical Controls

Prostar.

Resistant Species and Cultivars

Resistant cultivars of Kentucky bluegrass include: Adelphi and Cheri.

Management

Cultural Controls

Maintain moderate nitrogen fertility (1/2 lb. N/1000 sq.ft./month).
Maintain moderate to high levels of potash according to soil tests.
Reduce shade and increase air circulation to enhance drying of turf.
Reduce thatch thickness to 1/2 inch or less.

Yellow Ring

Hosts

Bluegrasses and bentgrasses.

Pathogen

Trechispora alnicola

Host Nutrition

Disease may be severe on lush, fast-growing turf that has received excess nitrogen (> 1/2 lb. N/1000 sq.ft./month).

Diagnosis

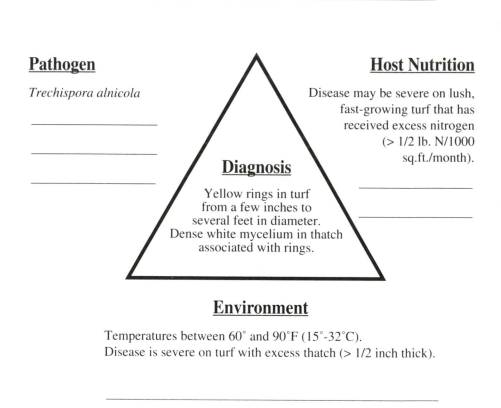

Yellow rings in turf from a few inches to several feet in diameter. Dense white mycelium in thatch associated with rings.

Environment

Temperatures between 60° and 90°F (15°-32°C).
Disease is severe on turf with excess thatch (> 1/2 inch thick).

Yellow Ring

Disease Forecasting and Pathogen Detection

A yellow ring forecaster and detection kits are not available.

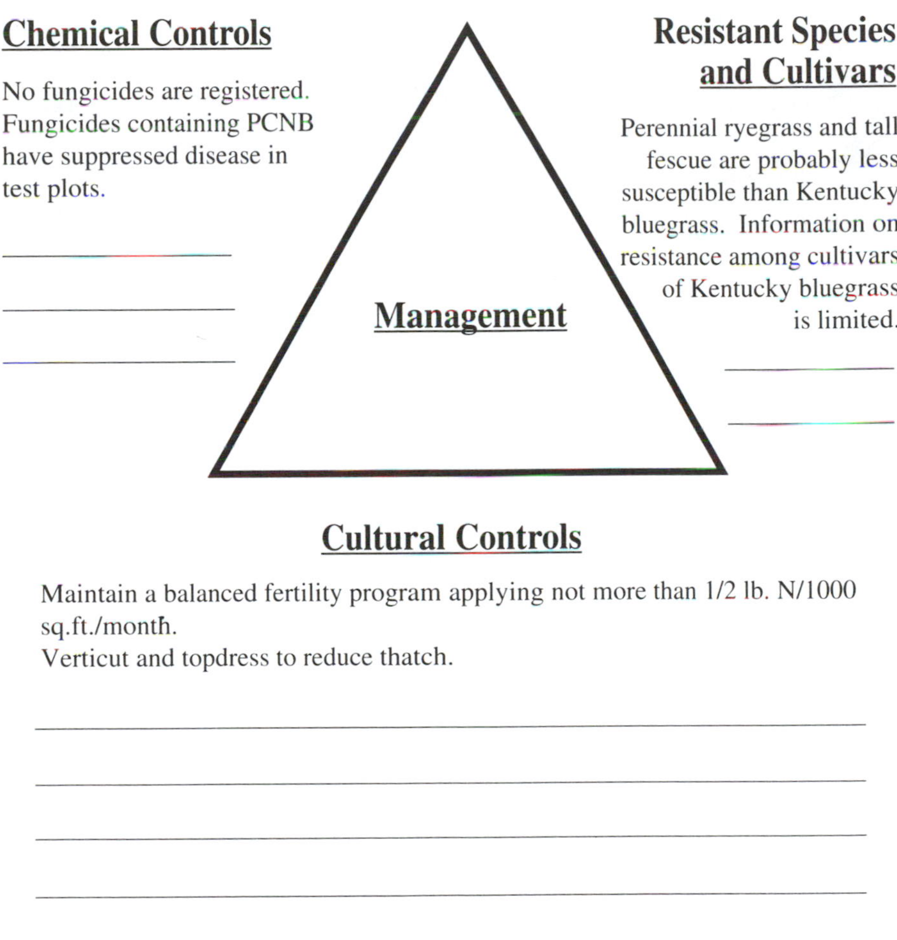

Chemical Controls

No fungicides are registered. Fungicides containing PCNB have suppressed disease in test plots.

Resistant Species and Cultivars

Perennial ryegrass and tall fescue are probably less susceptible than Kentucky bluegrass. Information on resistance among cultivars of Kentucky bluegrass is limited.

Cultural Controls

Maintain a balanced fertility program applying not more than 1/2 lb. N/1000 sq.ft./month.
Verticut and topdress to reduce thatch.

Chapter 3

Tall Fescue

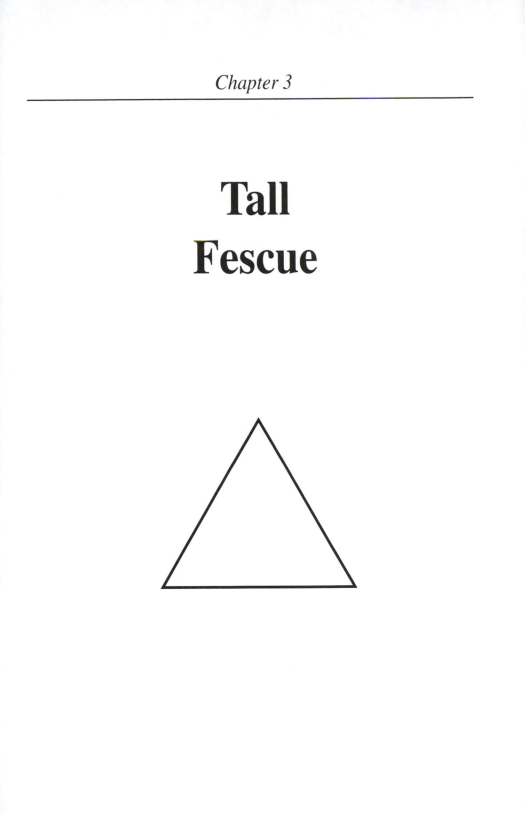

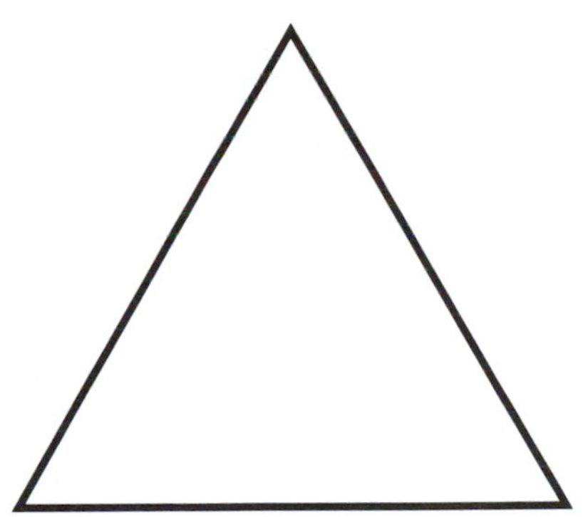

Brown Patch
(Rhizoctonia Blight)

Hosts

All common species of turfgrasses. Tall fescue is highly susceptible.

Pathogen

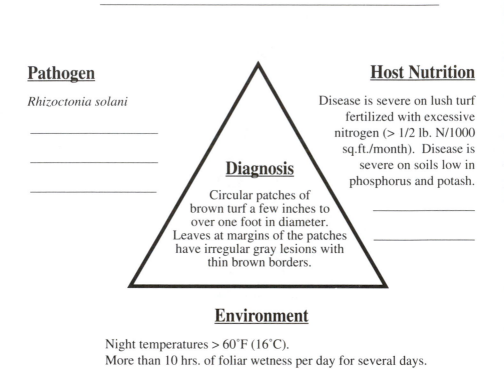

Rhizoctonia solani

Host Nutrition

Disease is severe on lush turf fertilized with excessive nitrogen (> 1/2 lb. N/1000 sq.ft./month). Disease is severe on soils low in phosphorus and potash.

Diagnosis

Circular patches of brown turf a few inches to over one foot in diameter. Leaves at margins of the patches have irregular gray lesions with thin brown borders.

Environment

Night temperatures > 60˚F (16˚C).
More than 10 hrs. of foliar wetness per day for several days.

Brown Patch

Disease Forecasting and Pathogen Detection

A brown patch forecaster and detection kits are available from Neogen Corp., 620 Lesher Pl., Lansing, MI 48912.

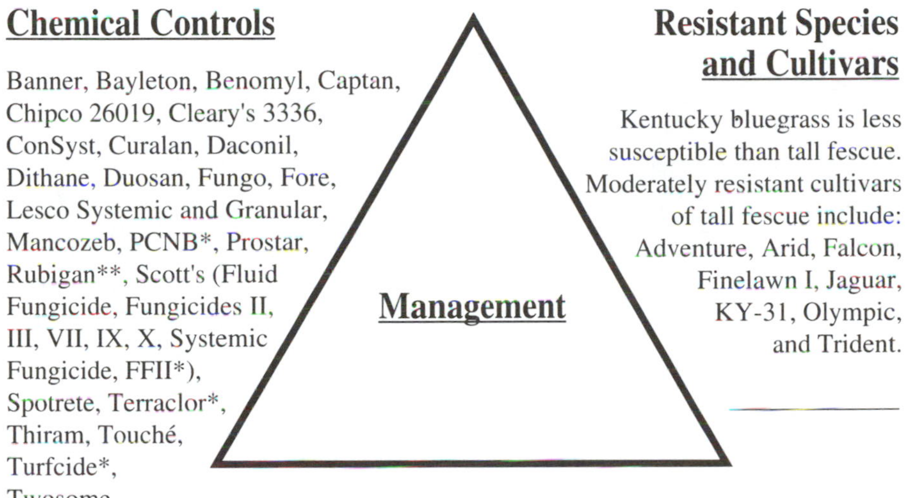

Chemical Controls

Banner, Bayleton, Benomyl, Captan, Chipco 26019, Cleary's 3336, ConSyst, Curalan, Daconil, Dithane, Duosan, Fungo, Fore, Lesco Systemic and Granular, Mancozeb, PCNB*, Prostar, Rubigan**, Scott's (Fluid Fungicide, Fungicides II, III, VII, IX, X, Systemic Fungicide, FFII*), Spotrete, Terraclor*, Thiram, Touché, Turfcide*, Twosome.

Resistant Species and Cultivars

Kentucky bluegrass is less susceptible than tall fescue. Moderately resistant cultivars of tall fescue include: Adventure, Arid, Falcon, Finelawn I, Jaguar, KY-31, Olympic, and Trident.

Cultural Controls

Maintain moderate nitrogen fertility (1/2 lb. N/1000 sq.ft./month).
Maintain moderate phosphorous and high potash according to soil tests.
Decrease shade and increase air circulation to enhance drying of turf.
Avoid irrigation in late afternoon and in evening prior to midnight.
Maintain thatch at 1/2 inch thick or less.
Raise mowing height to at least 2 inches.

* Avoid application to bentgrasses. ** May reduce populations of annual bluegrass.

Damping-off and Seed Rot

Hosts

All species of turfgrasses.

Pathogen

Species of *Pythium, Fusarium* or *Rhizoctonia*

Host Nutrition

Disease is severe on seed or seedlings fertilized excessively with nitrogen or on turf subjected to nutrient deficiency.

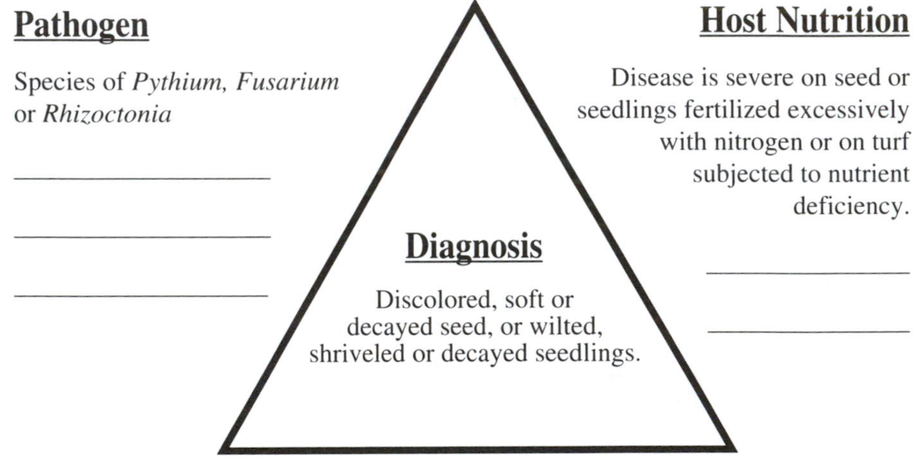

Diagnosis

Discolored, soft or decayed seed, or wilted, shriveled or decayed seedlings.

Environment

Temperatures too high (> 85°F, 30°C) or too low (< 60°F, 15°C) for optimum seedling development.
More than 10 hrs. of seed or seedling wetness per day for several days.
Excessive shade or overcrowding of seedlings.
Poor surface or subsurface drainage.

Damping-off and Seed Rot

Disease Forecasting and Pathogen Detection

A damping-off or seed rot forecaster is not available. Detection kits for ***Pythium*** and ***Rhizoctonia*** are available from Neogen Corp., 620 Lesher Pl., Lansing, MI 48912.

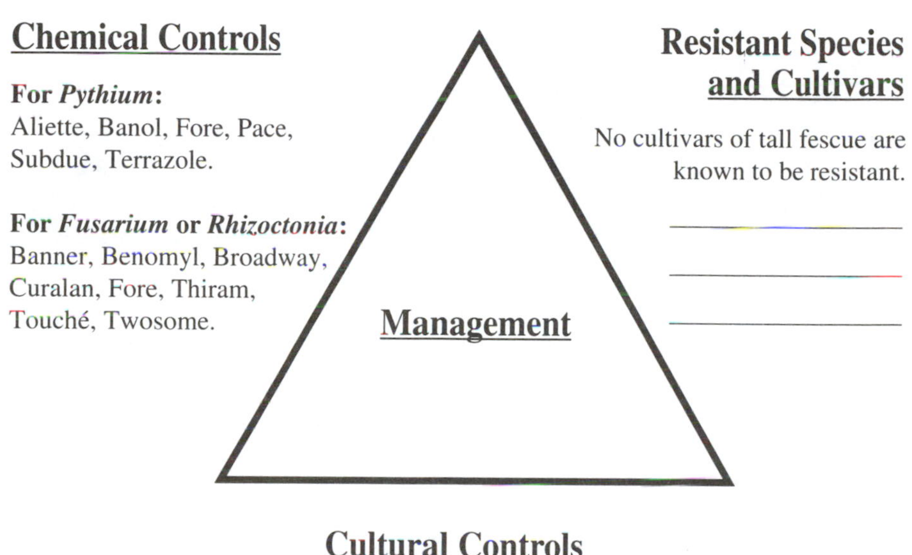

Chemical Controls

For *Pythium*:
Aliette, Banol, Fore, Pace, Subdue, Terrazole.

For *Fusarium* or *Rhizoctonia*:
Banner, Benomyl, Broadway, Curalan, Fore, Thiram, Touché, Twosome.

Resistant Species and Cultivars

No cultivars of tall fescue are known to be resistant.

Management

Cultural Controls

Incorporate 1-3 lbs. N/1000 sq.ft. in seed bed prior to seeding.
Incorporate phosphorous and potash according to soil tests.
Fertilize seedlings with 1/2 to 1 lb. N/1000 sq.ft./month.
Seed turf when day temperatures are between 60° and 80°F (15° and 27°C).
Use recommended seeding rate (7 to 9 lb. seed/1000 sq.ft.) for tall fescue.
Avoid high seeding rates.
Avoid irrigation in late afternoon and in evening prior to midnight.
Reduce shade and increase air circulation to enhance drying of turf.
Raise mowing height to at least 2 inches, and use light-weight equipment.
Improve surface and subsurface drainage.

* Some materials may be sold as seed treatments as well as foliar sprays.

Dollar Spot

Hosts

All common species of turfgrasses.

Pathogen

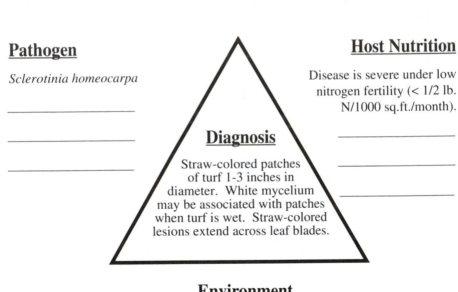

Sclerotinia homeocarpa

Host Nutrition

Disease is severe under low nitrogen fertility (< 1/2 lb. N/1000 sq.ft./month).

Diagnosis

Straw-colored patches of turf 1-3 inches in diameter. White mycelium may be associated with patches when turf is wet. Straw-colored lesions extend across leaf blades.

Environment

Night temperatures > 50°F (10°C) and day temperatures < 90°F (32°C).
More than 10 hrs. of leaf wetness per day for several days.
Disease is severe on turf subjected to drought stress.

Dollar Spot

Disease Forecasting and Pathogen Detection

A dollar spot forecaster is available from Pest Management Supply, P.O. Box 938, Amherst, MA 01004.
Detection kits are available from Neogen Corp., 620 Lesher Pl., Lansing, MI 48912.

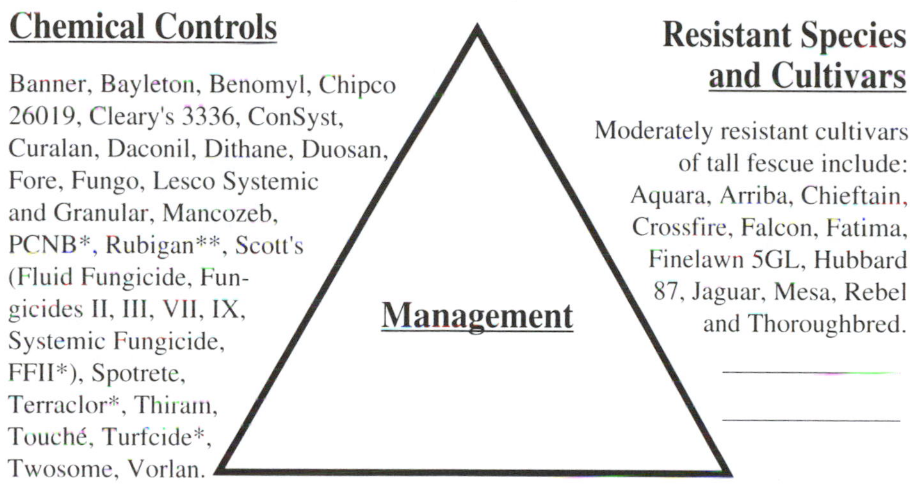

Chemical Controls

Banner, Bayleton, Benomyl, Chipco 26019, Cleary's 3336, ConSyst, Curalan, Daconil, Dithane, Duosan, Fore, Fungo, Lesco Systemic and Granular, Mancozeb, PCNB*, Rubigan**, Scott's (Fluid Fungicide, Fungicides II, III, VII, IX, Systemic Fungicide, FFII*), Spotrete, Terraclor*, Thiram, Touché, Turfcide*, Twosome, Vorlan.

Resistant Species and Cultivars

Moderately resistant cultivars of tall fescue include: Aquara, Arriba, Chieftain, Crossfire, Falcon, Fatima, Finelawn 5GL, Hubbard 87, Jaguar, Mesa, Rebel and Thoroughbred.

Cultural Controls

Applications of 1/2 to 1 lb. of N/1000 sq. ft. every 2-4 weeks will reduce severity of dollar spot.
Maintain moderate to high levels of soil potassium as determined by soil-tests.
Limit thatch to 1/2 inch or less.
Decrease shade and increase air circulation to enhance drying of turf.
Avoid irrigation in late afternoon and in evening prior to midnight.
Avoid drought stress.
Avoid mowing at height < 2 inches.

* Avoid application to bentgrasses. ** May reduce populations of annual bluegrass.

Fairy Ring

Hosts

All turfgrasses.

Pathogen

Several species of "mushroom-forming" fungi.

Host Nutrition

High nitrogen fertility (> 1/2 lb. N/1000 sq.ft./month) may increase disease severity. Low nitrogen may increase the frequency of occurrence of fairy ring.

Diagnosis

Circles or archs of mushrooms or wilted, dead or dark green turf. White mats of fungal mycelium may be found in thatch or soil associated with circles or archs.

Environment

Light to moderate textured soils.
Soil pH of 5 to 7.5.
Low to moderate soil mosture.

Fairy Ring

Disease Forecasting and Pathogen Detection

A fairy ring forecaster and detection kits are not available.

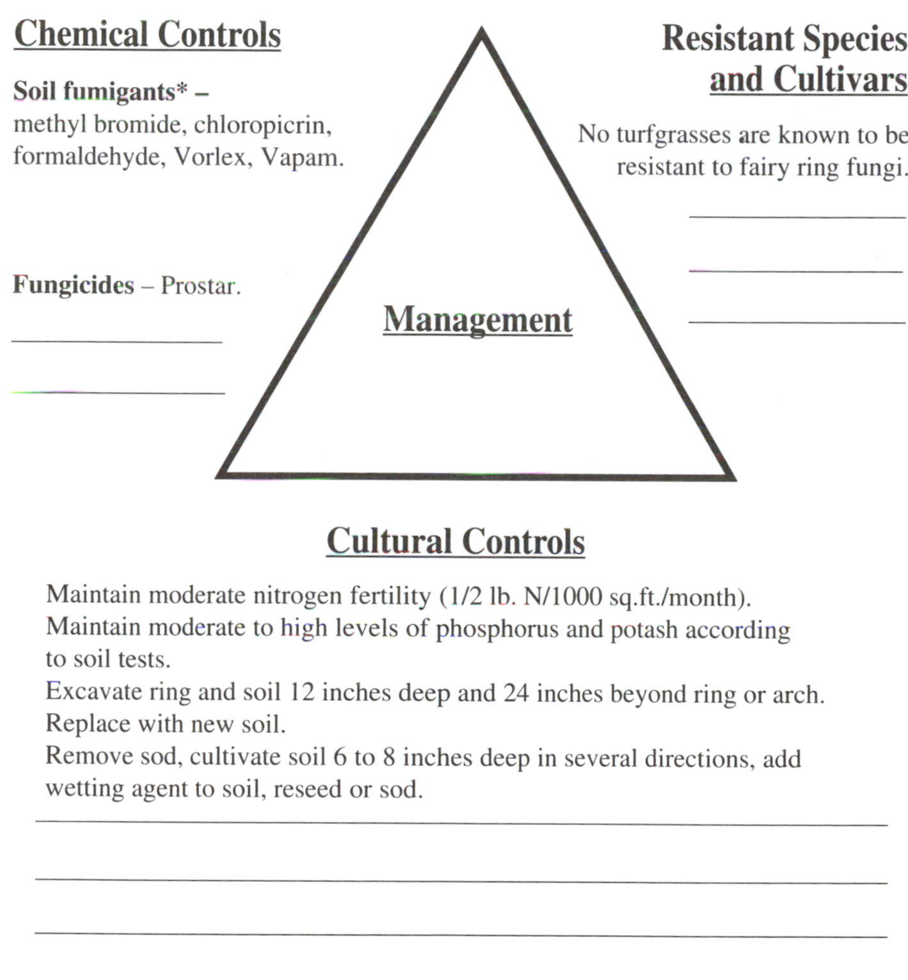

Chemical Controls

Soil fumigants* –
methyl bromide, chloropicrin, formaldehyde, Vorlex, Vapam.

Fungicides – Prostar.

Resistant Species and Cultivars

No turfgrasses are known to be resistant to fairy ring fungi.

Cultural Controls

Maintain moderate nitrogen fertility (1/2 lb. N/1000 sq.ft./month).
Maintain moderate to high levels of phosphorus and potash according to soil tests.
Excavate ring and soil 12 inches deep and 24 inches beyond ring or arch. Replace with new soil.
Remove sod, cultivate soil 6 to 8 inches deep in several directions, add wetting agent to soil, reseed or sod.

* These chemicals are highly toxic to turfgrasses, animals and other life forms.

Fusarium Patch

Hosts

All cool season turfgrasses. Annual bluegrass, bentgrasses, and perennial ryegrass are particularly susceptible.

Pathogen

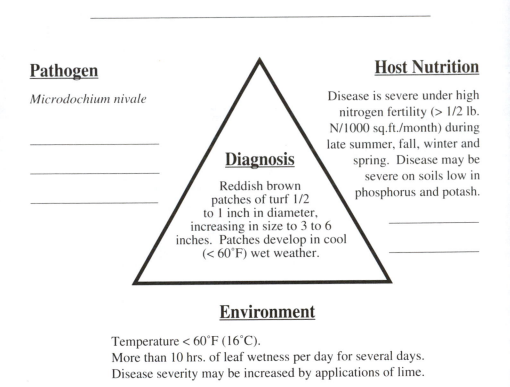

Microdochium nivale

Host Nutrition

Disease is severe under high nitrogen fertility (> 1/2 lb. N/1000 sq.ft./month) during late summer, fall, winter and spring. Disease may be severe on soils low in phosphorus and potash.

Diagnosis

Reddish brown patches of turf 1/2 to 1 inch in diameter, increasing in size to 3 to 6 inches. Patches develop in cool (< 60˚F) wet weather.

Environment

Temperature < 60˚F (16˚C).
More than 10 hrs. of leaf wetness per day for several days.
Disease severity may be increased by applications of lime.

Fusarium Patch

Disease Forecasting and Pathogen Detection

A Fusarium patch forecaster and detection kits are not available.

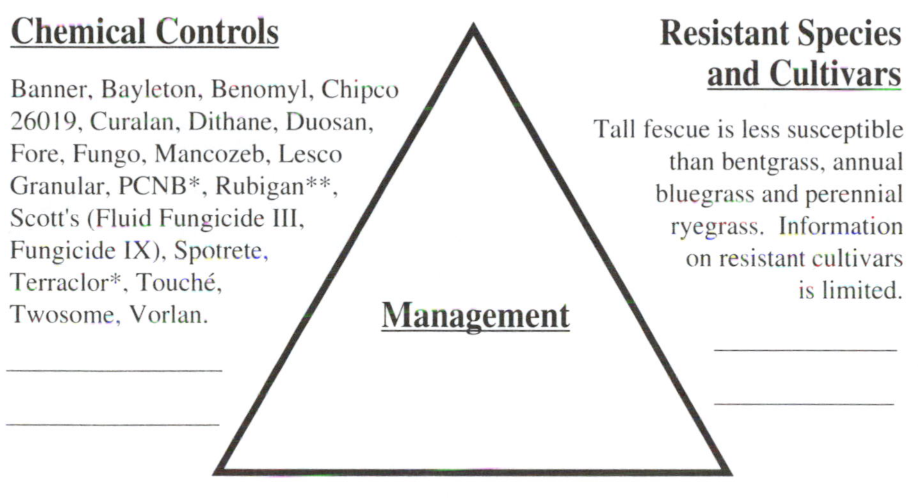

Chemical Controls

Banner, Bayleton, Benomyl, Chipco 26019, Curalan, Dithane, Duosan, Fore, Fungo, Mancozeb, Lesco Granular, PCNB*, Rubigan**, Scott's (Fluid Fungicide III, Fungicide IX), Spotrete, Terraclor*, Touché, Twosome, Vorlan.

Resistant Species and Cultivars

Tall fescue is less susceptible than bentgrass, annual bluegrass and perennial ryegrass. Information on resistant cultivars is limited.

Management

Cultural Controls

Avoid high nitrogen (> 1/2 lb. N/1000 sq.ft./month) in late summer and early fall.
Maintain moderate to high levels of phosphorous and potash according to soil tests.
Decrease shade and increase air circulation to enhance drying of turf.
Avoid applications of lime if possible.
Avoid irrigation in late afternoon and in evening prior to midnight.
Avoid mowing at height < 2 inches.

* Avoid application to actively growing bentgrass. ** May reduce populations of annual bluegrass.

Leaf Spot

Hosts

Bluegrasses, bentgrasses, fescues and perennial ryegrass.

Pathogen

Bipolaris sorokiniana

Host Nutrition

Disease is severe under high nitrogen fertility (> 1/2 lb. N/1000 sq.ft./month).

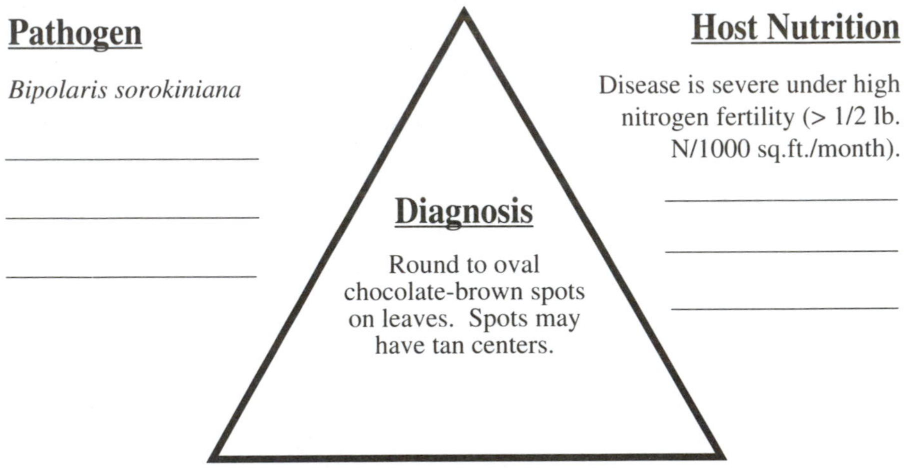

Diagnosis

Round to oval chocolate-brown spots on leaves. Spots may have tan centers.

Environment

Temperatures of 77°-95°F (25°-35°C).
Disease severity increases with increases in temperature.
More than 10 hrs. of leaf wetness per day for several days.

Leaf Spot

Disease Forecasting and Pathogen Detection

A leaf spot forecaster and detection kits are not available.

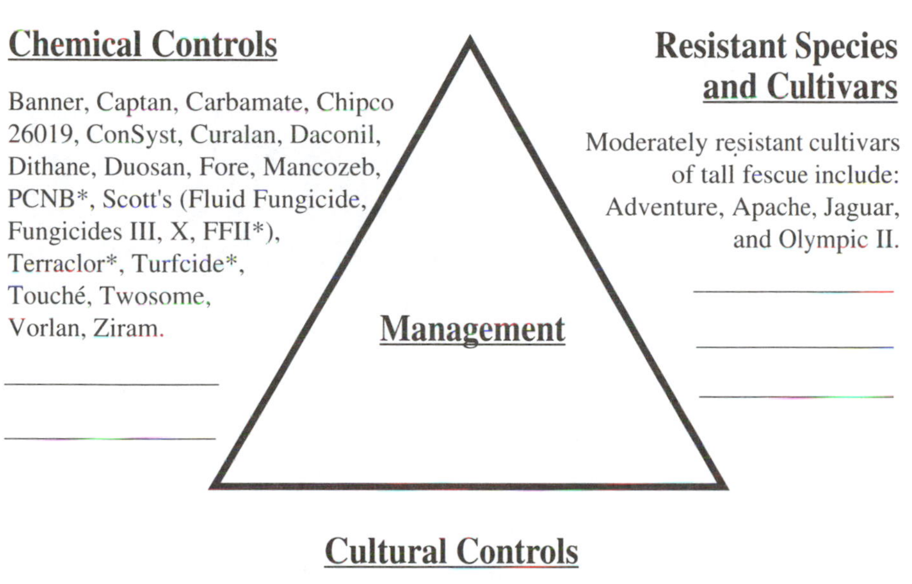

Chemical Controls

Banner, Captan, Carbamate, Chipco 26019, ConSyst, Curalan, Daconil, Dithane, Duosan, Fore, Mancozeb, PCNB*, Scott's (Fluid Fungicide, Fungicides III, X, FFII*), Terraclor*, Turfcide*, Touché, Twosome, Vorlan, Ziram.

Management

Resistant Species and Cultivars

Moderately resistant cultivars of tall fescue include: Adventure, Apache, Jaguar, and Olympic II.

Cultural Controls

Apply moderate amounts of nitrogen during summer (1/4-1/2 lb. N/1000 sq.ft./month).
Maintain moderate to high levels of soil P and K.
Decrease shade and increase air circulation to enhance drying of turf.
Avoid irrigation in late afternoon and in evening prior to midnight.
Limit thatch to 1/2 inch or less.
Raise mowing height to at least 2 inches.
Use light-weight mowing equipment to reduce stress.

* Avoid application to bentgrasses.

Melting-Out

Hosts

Bluegrasses, ryegrasses and tall fescue.

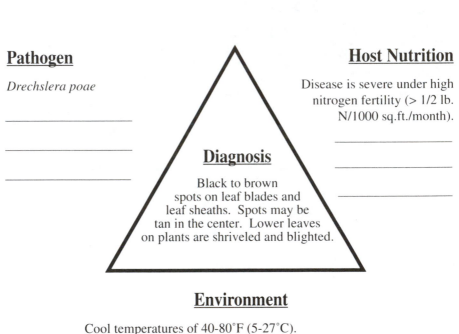

Pathogen

Drechslera poae

Host Nutrition

Disease is severe under high nitrogen fertility (> 1/2 lb. N/1000 sq.ft./month).

Diagnosis

Black to brown spots on leaf blades and leaf sheaths. Spots may be tan in the center. Lower leaves on plants are shriveled and blighted.

Environment

Cool temperatures of 40-80°F (5-27°C).
More than 10 hrs. of leaf wetness per day for several days.
Mowing height < 2 inches.

Melting-Out

Disease Forecasting and Pathogen Detection

A melting-out forecaster and detection kits are not available.

Chemical Controls

Banner, Captan, Chipco 26019, ConSyst, Curalan, Daconil, Dithane, Fore, Mancozeb, Scott's (Fluid Fungicide, Fungicides III), Terraclor*, Turfcide*, Touché, Twosome, Vorlan.

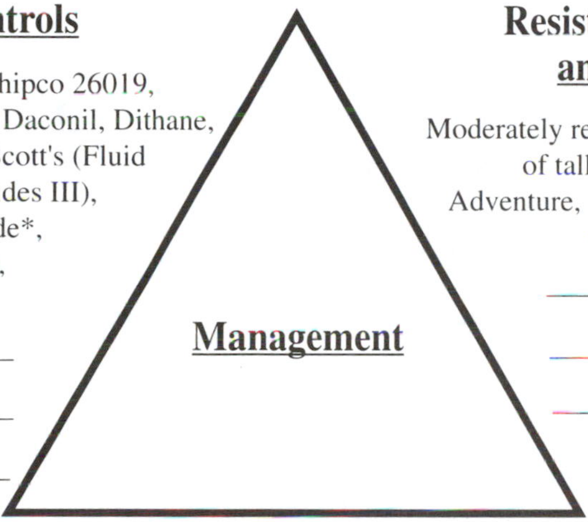

Resistant Species and Cultivars

Moderately resistant cultivars of tall fescue include: Adventure, Apache, Jaguar, and Olympic II.

Management

Cultural Controls

Fertilize with low to moderate levels of N during spring, summer and fall (1/4-1/2 lb. N/1000 sq.ft./month).
Decrease shade and increase air circulation to enhance drying of turf.
Avoid irrigation in late afternoon and in evening prior to midnight.
Raise mowing height to at least 2 inches.
Use light-weight mowing equipment to avoid stress on turf.
Limit thatch thickness to 1/4 inch or less.

* Avoid application to bentgrasses.

Nematodes

Hosts

All common species of turfgrasses.

Pathogen

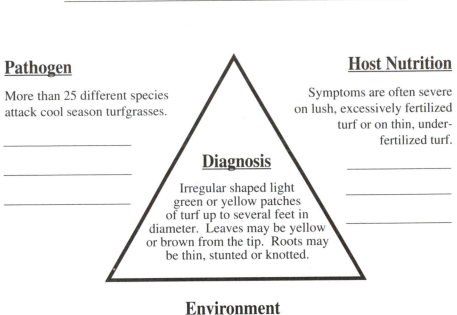

More than 25 different species attack cool season turfgrasses.

Host Nutrition

Symptoms are often severe on lush, excessively fertilized turf or on thin, under-fertilized turf.

Diagnosis

Irregular shaped light green or yellow patches of turf up to several feet in diameter. Leaves may be yellow or brown from the tip. Roots may be thin, stunted or knotted.

Environment

Soil temperatures > 40°F (5°C).
Symptoms are often severe on turf growing in sandy, light-textured soils.
Symptoms may be enhanced by drought and high temperatures (> 80°F, 26°C).

Nematodes

Disease Forecasting and Pathogen Detection

A nematode forecaster and detection kits are not available.

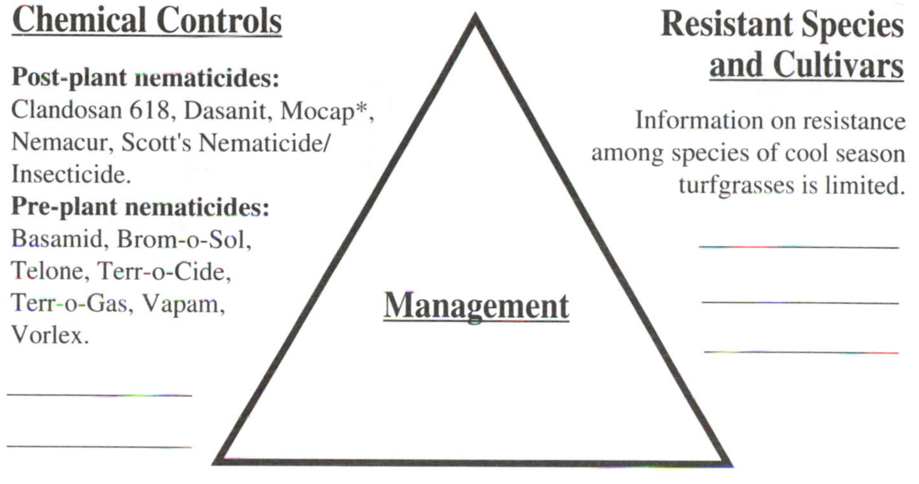

Chemical Controls

Post-plant nematicides:
Clandosan 618, Dasanit, Mocap*, Nemacur, Scott's Nematicide/ Insecticide.

Pre-plant nematicides:
Basamid, Brom-o-Sol, Telone, Terr-o-Cide, Terr-o-Gas, Vapam, Vorlex.

Management

Resistant Species and Cultivars

Information on resistance among species of cool season turfgrasses is limited.

Cultural Controls

Maintain a balanced fertility program.
Apply 1/2 lb. N/1000 sq.ft./month during spring and summer.
Maintain moderate to high levels of phosphorus and potash according to soil tests.
Have soil analyzed for nematodes prior to seeding or sodding.
Use sod that is nematode-free.

* Avoid applications to bentgrasses.

Net Blotch

Hosts

Fescues, bluegrass and ryegrasses.

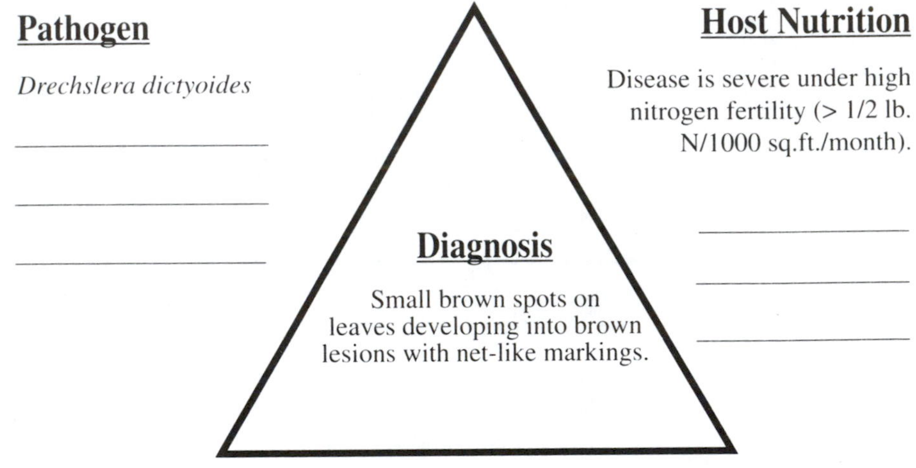

Pathogen

Drechslera dictyoides

Host Nutrition

Disease is severe under high nitrogen fertility (> 1/2 lb. N/1000 sq.ft./month).

Diagnosis

Small brown spots on leaves developing into brown lesions with net-like markings.

Environment

Cool temperatures of 40-80°F (5-27°C).
More than 10 hrs. of leaf wetness per day for several days.
Mowing height < 2 inches.

Net Blotch

Disease Forecasting and Pathogen Detection

A net blotch forecaster and detection kits are not available.

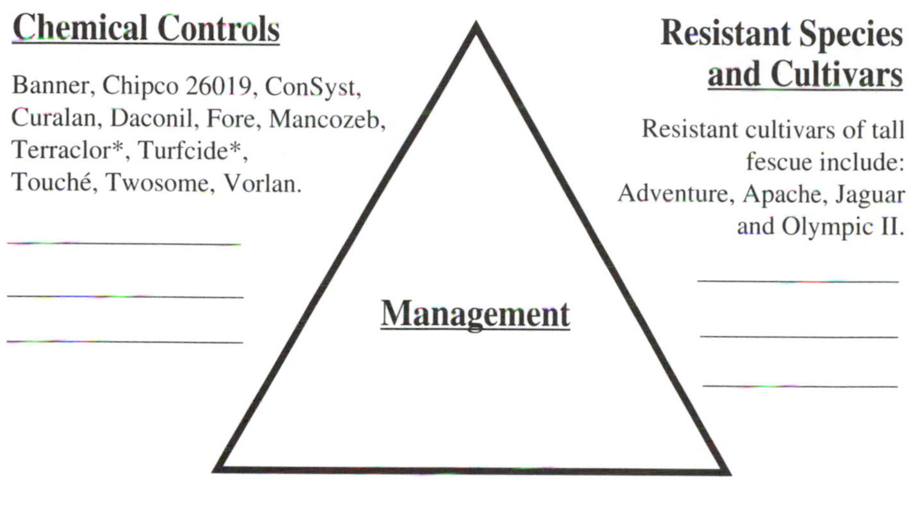

Chemical Controls
Banner, Chipco 26019, ConSyst,
Curalan, Daconil, Fore, Mancozeb,
Terraclor*, Turfcide*,
Touché, Twosome, Vorlan.
Management
Resistant Species
and Cultivars
Resistant cultivars of tall
fescue include:
Adventure, Apache, Jaguar
and Olympic II.

Cultural Controls

Fertilize with low to moderate levels of N during spring, summer and fall (1/4-1/2 lb. N/1000 sq.ft./month).
Decrease shade and increase air circulation to enhance drying of turf.
Avoid irrigation in late afternoon and in evening prior to midnight.
Raise mowing height to at least 2 inches.
Use light-weight mowing equipment to avoid stress on turf.

* Avoid applications to bentgrasses.

Pink Snow Mold

Hosts

All cool season grasses. Bentgrasses, annual bluegrass, and perennial ryegrass are particularly susceptible.

Pathogen

Microdochium nivale (same pathogen causes Fusarium Patch)

Host Nutrition

Fast-growing, lush turf, that receives high nitrogen and low potash in late fall is particularly susceptible.

Diagnosis

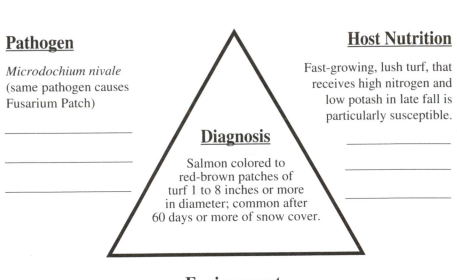

Salmon colored to red-brown patches of turf 1 to 8 inches or more in diameter; common after 60 days or more of snow cover.

Environment

Disease is common after at least 60 days of snow cover, but pathogen can infect turf in absence of snow (see Fusarium Patch).
Disease is particularly severe when snow covers unfrozen ground.

Pink Snow Mold

Disease Forecasting and Pathogen Detection

A pink snow mold forecaster and detection kits are not available.

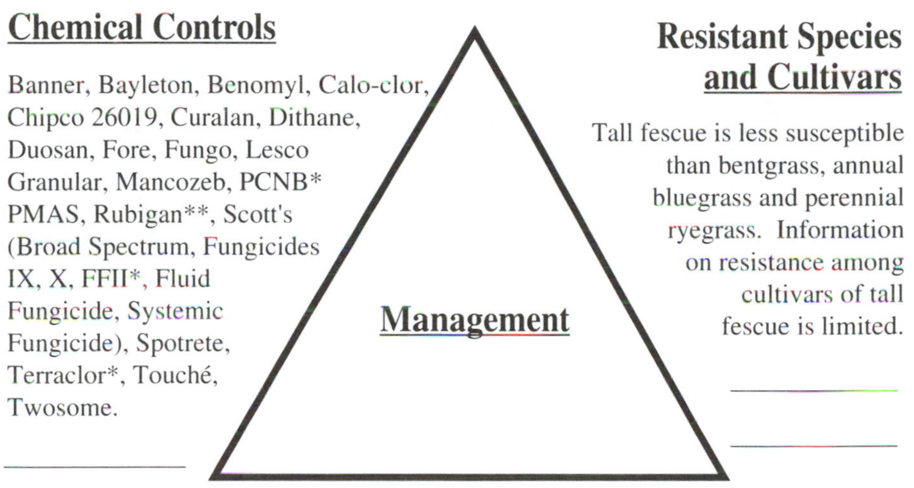

Chemical Controls

Banner, Bayleton, Benomyl, Calo-clor, Chipco 26019, Curalan, Dithane, Duosan, Fore, Fungo, Lesco Granular, Mancozeb, PCNB* PMAS, Rubigan**, Scott's (Broad Spectrum, Fungicides IX, X, FFII*, Fluid Fungicide, Systemic Fungicide), Spotrete, Terraclor*, Touché, Twosome.

Management

Resistant Species and Cultivars

Tall fescue is less susceptible than bentgrass, annual bluegrass and perennial ryegrass. Information on resistance among cultivars of tall fescue is limited.

Cultural Controls

Maintain moderate nitrogen fertility (1/2 lb. N/1000 sq.ft./month) during late summer and fall.
Maintain high potash levels according to soil tests.
Use snow fence, shrubs or knolls as windbreaks to prevent excess snow from accumulating.
Prevent snow compaction by machinery or skiers.
Melt snow in spring with organic fertilizers.
Physically remove snow in spring.
Follow controls for Fusarium Patch after snow melt.

* Avoid applications to bentgrasses. **May reduce populations of annual bluegrass.

Pythium Blight

Hosts

All cool season grasses. Annual bluegrass and perennial ryegrass are particularly susceptible.

Pathogen

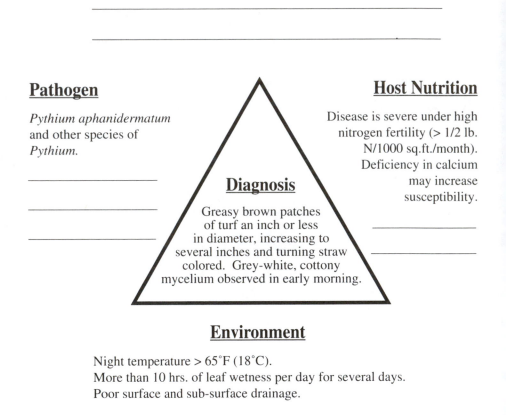

Pythium aphanidermatum and other species of *Pythium.*

Host Nutrition

Disease is severe under high nitrogen fertility (> 1/2 lb. N/1000 sq.ft./month). Deficiency in calcium may increase susceptibility.

Diagnosis

Greasy brown patches of turf an inch or less in diameter, increasing to several inches and turning straw colored. Grey-white, cottony mycelium observed in early morning.

Environment

Night temperature > 65°F (18°C).
More than 10 hrs. of leaf wetness per day for several days.
Poor surface and sub-surface drainage.

Pythium Blight

Disease Forecasting and Pathogen Detection

Pythium blight forecasters are available from Pest Management Supply, P.O. Box 936, Amherst, MA 01004 or Neogen Corp., 620 Lesher Pl., Lansing, MI 48912. Detection kits are available from Neogen Corp.

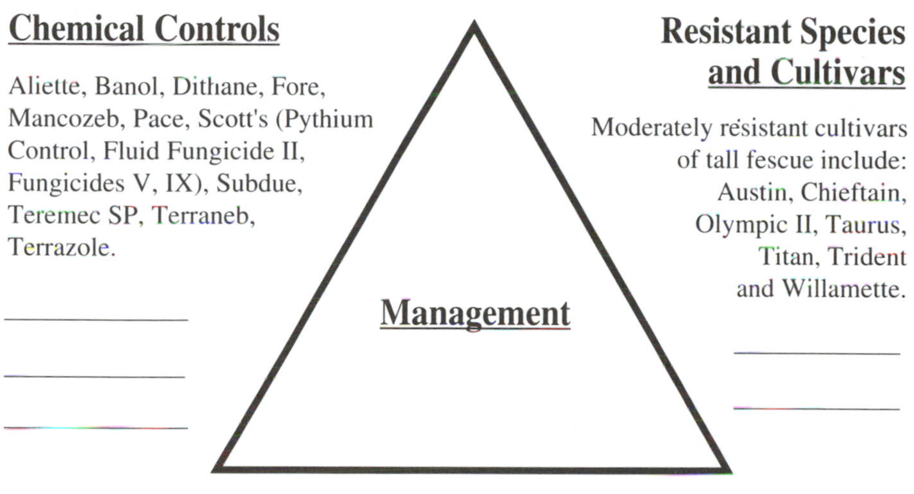

Chemical Controls

Aliette, Banol, Dithane, Fore, Mancozeb, Pace, Scott's (Pythium Control, Fluid Fungicide II, Fungicides V, IX), Subdue, Teremec SP, Terraneb, Terrazole.

Resistant Species and Cultivars

Moderately resistant cultivars of tall fescue include: Austin, Chieftain, Olympic II, Taurus, Titan, Trident and Willamette.

Cultural Controls

Maintain moderate nitrogen fertility (1/2 lb. N/1000 sq.ft./month).
Maintain optimum plant calcium levels.
Decrease shade and increase air circulation to enhance drying of turf.
Improve surface and subsurface drainage.
Avoid mowing susceptible areas when turf is wet, particularly when night temperatures are > 70°F (21°C).

Tall Fescue

Pythium Root Rot

Hosts

All species of cool season grasses.

Pathogen

Pythium aphanidermatum, P. graminicola, P. myriotylum, plus other *Pythium* species.

Host Nutrition

Disease may be severe if nitrogen is deficient or excessive.

Diagnosis

Irregular yellow brown to straw-colored patches of turf 1-6 inches or more in diameter. Roots stunted, thin, grey-white to brown.

Environment

Cool (32°-50°F, 0-10°C) or warm (70°-90°F, 21-32°C) soil temperatures.*
High soil mosture.
Poor surface or subsurface drainage.
Conditions unfavorable for carbohydrate development by leaves – low light, low mowing height, excessive wear.

* Some *Pythium* species are favored by cool soils, other species by warm soils.

Pythium Root Rot

Disease Forecasting and Pathogen Detection

A Pythium root rot forecaster is not available. Detection kits are available from Neogen Corp., 620 Lesher Pl., Lansing, MI 48912.

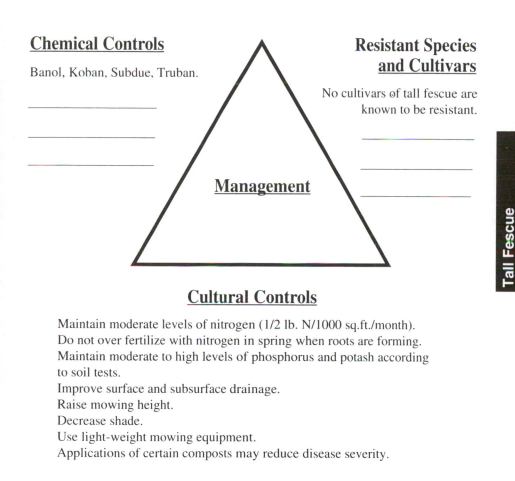

Chemical Controls

Banol, Koban, Subdue, Truban.

Resistant Species and Cultivars

No cultivars of tall fescue are known to be resistant.

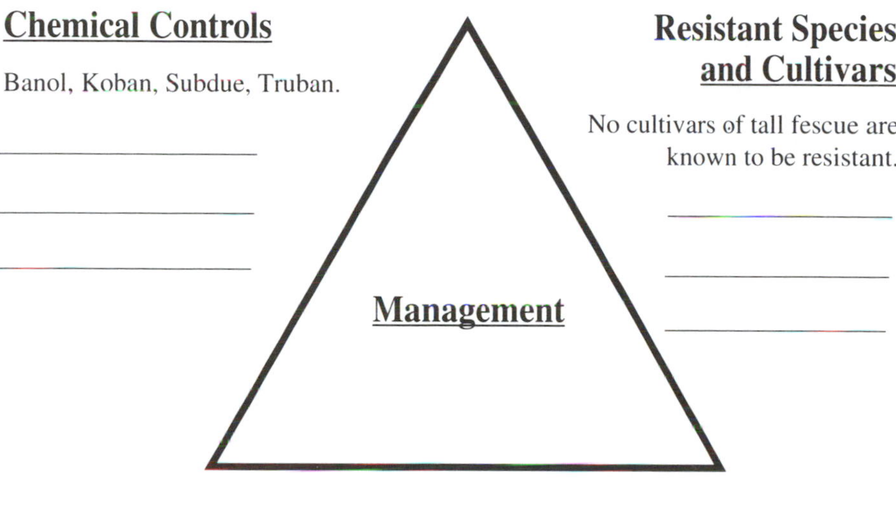

Cultural Controls

Maintain moderate levels of nitrogen (1/2 lb. N/1000 sq.ft./month).
Do not over fertilize with nitrogen in spring when roots are forming.
Maintain moderate to high levels of phosphorus and potash according to soil tests.
Improve surface and subsurface drainage.
Raise mowing height.
Decrease shade.
Use light-weight mowing equipment.
Applications of certain composts may reduce disease severity.

Rhizoctonia Leaf and Sheath Spot

Hosts

Bentgrass, bluegrasses, perennial ryegrass and tall fescue.

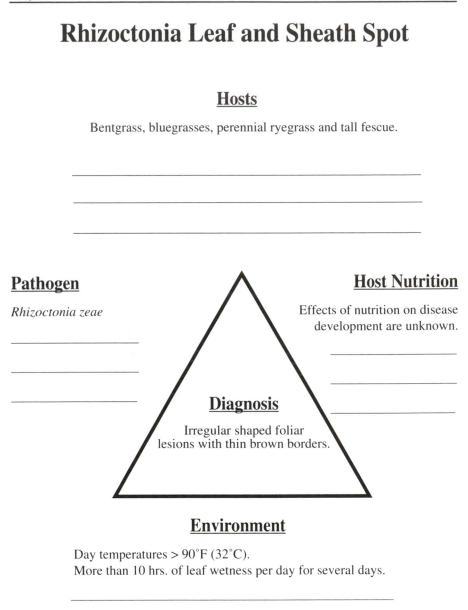

Pathogen

Rhizoctonia zeae

Host Nutrition

Effects of nutrition on disease development are unknown.

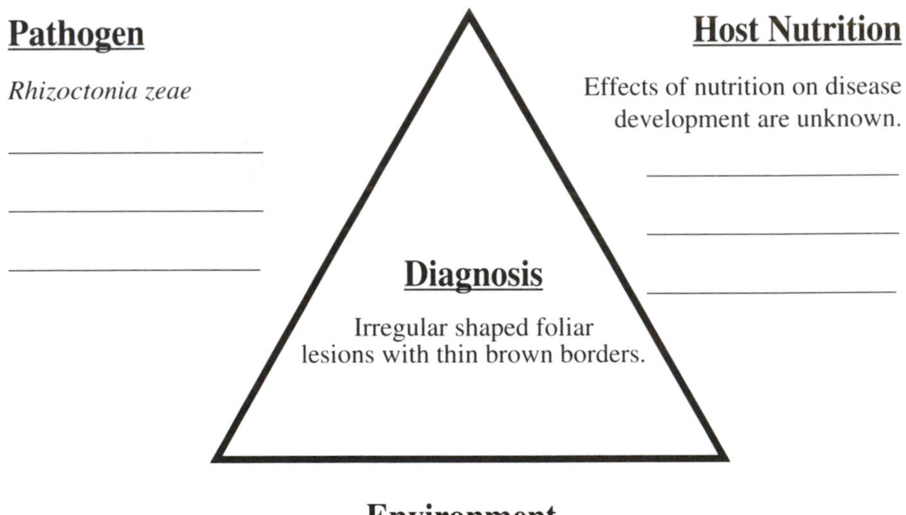

Diagnosis

Irregular shaped foliar lesions with thin brown borders.

Environment

Day temperatures > 90°F (32°C).
More than 10 hrs. of leaf wetness per day for several days.

Rhizoctonia Leaf and Sheath Spot

Disease Forecasting and Pathogen Detection

A leaf and sheath spot forecaster and detection kits are not available.

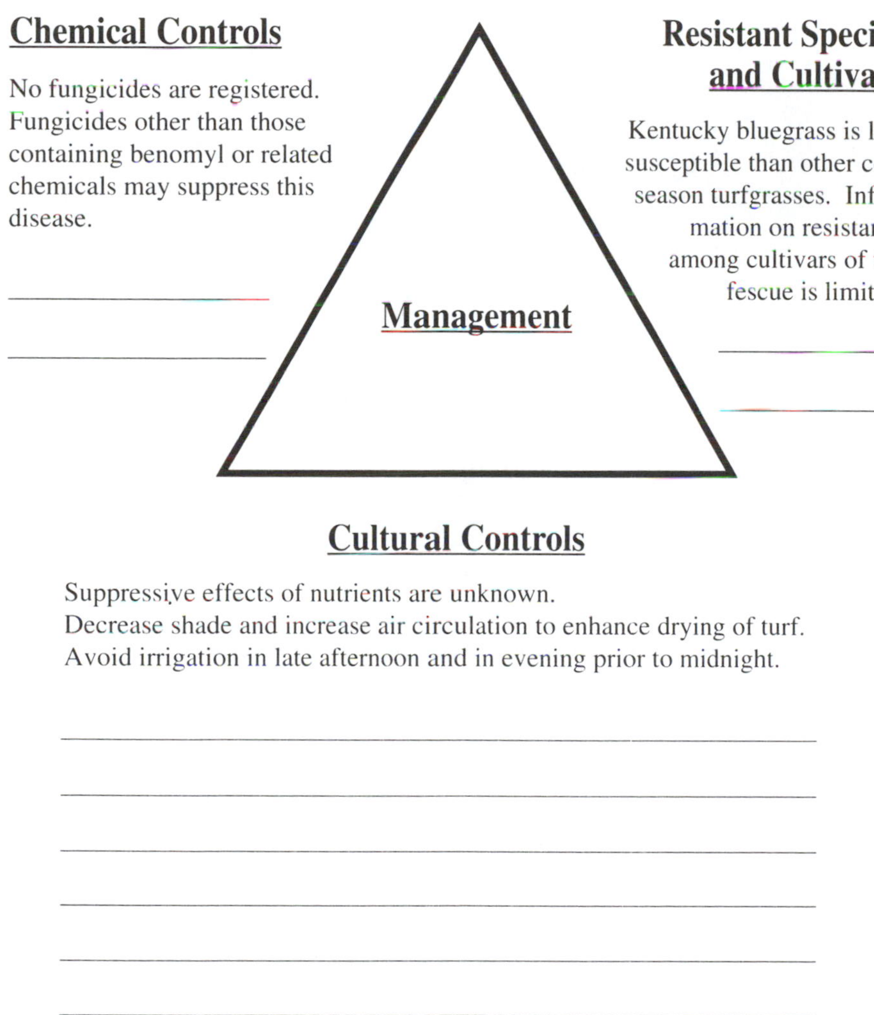

Chemical Controls

No fungicides are registered. Fungicides other than those containing benomyl or related chemicals may suppress this disease.

Resistant Species and Cultivars

Kentucky bluegrass is less susceptible than other cool season turfgrasses. Information on resistance among cultivars of tall fescue is limited.

Cultural Controls

Suppressive effects of nutrients are unknown.
Decrease shade and increase air circulation to enhance drying of turf.
Avoid irrigation in late afternoon and in evening prior to midnight.

Rust

Hosts

All common species of cool season turfgrasses.

Pathogen

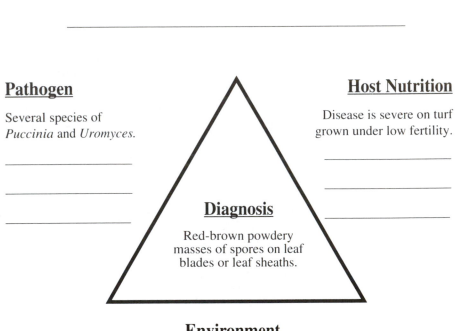

Several species of *Puccinia* and *Uromyces*.

Host Nutrition

Disease is severe on turf grown under low fertility.

Diagnosis

Red-brown powdery masses of spores on leaf blades or leaf sheaths.

Environment

Temperatures of 68°-86°F (20°-30°C).
Disease is severe on turf subjected to drought stress, low mowing, shade or poor air circulation.

Rust

Disease Forecasting and Pathogen Detection

A rust forecaster and detection kits are not available.

Chemical Controls

Banner, Bayleton, Captan, Carbamate, Cleary's 3336, ConSyst, Daconil, Dithane, Duosan, Fore, Mancozeb, Rubigan*, Scott's (Fluid Fungicide III, Fungicide VII, FFII**), Twosome, Ziram.

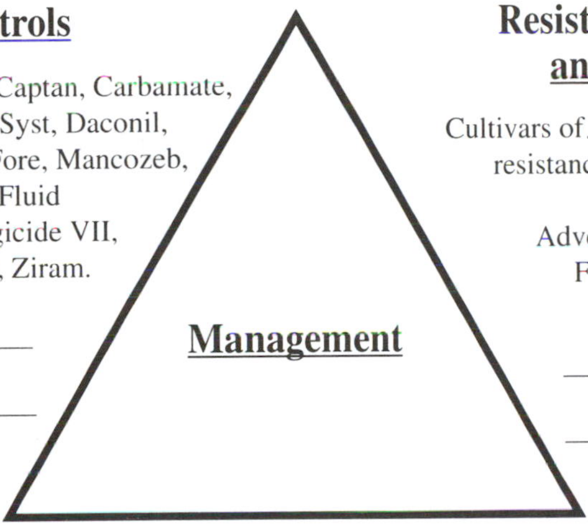

Resistant Species and Cultivars

Cultivars of.tall fescue with resistance to crown rust include: Adventure, Apache, Falcon, Mustang and Olympic.

Management

Cultural Controls

Maintain moderate and balanced fertility throughout the growing season.
Reduce shade and increase air circulation.
Increase mowing height to at least 2 inches.
Avoid drought stress.
Avoid irrigation in late afternoon and in early evening prior to midnight.

* May reduce populations of annual bluegrass. **Avoid application to bentgrasses.

Stripe Smut

Hosts

Bluegrasses, bentgrasses, perennial ryegrass and tall fescue. Kentucky bluegrass is more susceptible than other cool season grasses.

Pathogen

Ustilago striiformis

Host Nutrition

Disease is severe on lush overfertilized or on thin underfertilized turf.

Diagnosis

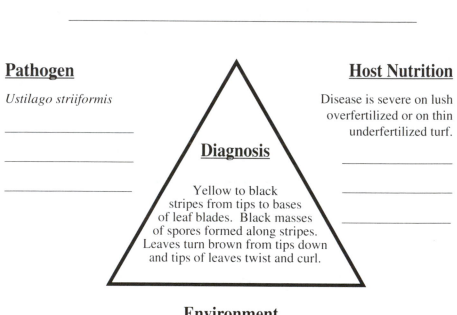

Yellow to black stripes from tips to bases of leaf blades. Black masses of spores formed along stripes. Leaves turn brown from tips down and tips of leaves twist and curl.

Environment

Infection occurs at 50°- 68°F (10°-20°C).
Severe symptoms evident during drought and temperatures > 75°F (24°C).
Symptoms are often more severe on acid soils and on turf with excessive thatch (> 1/2 inch thick).

Stripe Smut

Disease Forecasting and Pathogen Detection

A stripe smut forecaster and detection kits are not available.

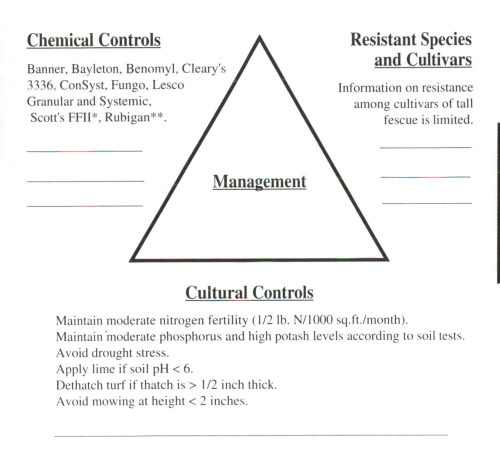

Chemical Controls

Banner, Bayleton, Benomyl, Cleary's 3336, ConSyst, Fungo, Lesco Granular and Systemic, Scott's FFII*, Rubigan**.

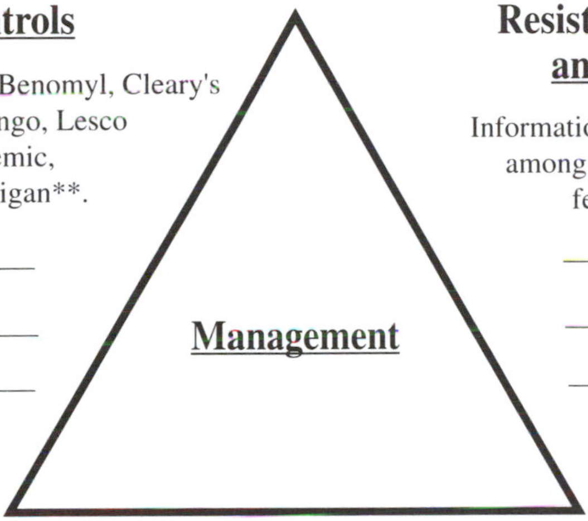

Resistant Species and Cultivars

Information on resistance among cultivars of tall fescue is limited.

Management

Cultural Controls

Maintain moderate nitrogen fertility (1/2 lb. N/1000 sq.ft./month).
Maintain moderate phosphorus and high potash levels according to soil tests.
Avoid drought stress.
Apply lime if soil pH < 6.
Dethatch turf if thatch is > 1/2 inch thick.
Avoid mowing at height < 2 inches.

* Avoid applications to bentgrasses. ** May reduce populations of annual bluegrass.

Typhula Blight
(Gray Snow Mold)

Hosts

All cool season turfgrasses. Bentgrasses, annual bluegrass and perennial ryegrass are particularly susceptible.

Pathogen

Typhula incarnata and *T. ishikariensis*

Host Nutrition

Disease is severe on lush, fast-growing turf that is covered with snow. High nitrogen and low potash in fall can predispose turf to severe damage.

Diagnosis

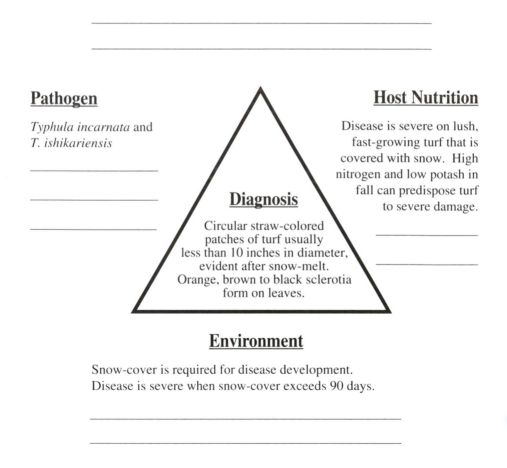

Circular straw-colored patches of turf usually less than 10 inches in diameter, evident after snow-melt. Orange, brown to black sclerotia form on leaves.

Environment

Snow-cover is required for disease development. Disease is severe when snow-cover exceeds 90 days.

Typhula Blight

Disease Forecasting and Pathogen Detection

A Typhula blight forecaster and detection kits are not available.

Chemical Controls

Banner, Bayleton, Calo-clor, Calo-gran, Chipco 26019, Curalan, Daconil, Lesco Granular, PCNB*, PMAS, Prostar, Rubigan**, Scott's (Broad Spectrum, Fungicides V, IX, FFII*), Spotrete, Terraclor*, Teremec SP, Thiram, Touché, Turfcide*, Twosome.

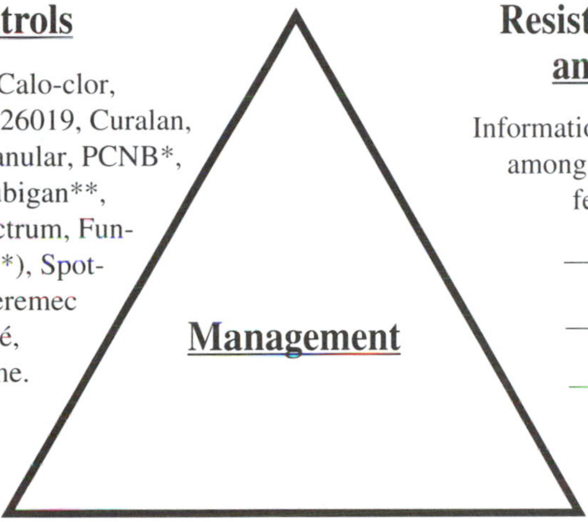

Resistant Species and Cultivars

Information on resistance among cultivars of tall fescue is limited.

Management

Cultural Controls

Avoid a fertility program that results in lush, fast-growing turf in late fall and winter.
Maintain high potash levels according to soil tests.
Use snow fence, hedges or knolls to prevent snow from accumulating excessively on turf.
Use dark-colored organic fertilizers or composts to melt snow in spring.
Physically remove snow in spring.
Prevent compaction of snow during winter.

*Avoid application to actively growing bentgrasses. **May reduce populations of annual bluegrass.

White Patch

(White Blight)

Hosts

Fescues, bluegrasses and creeping bentgrass.

Pathogen

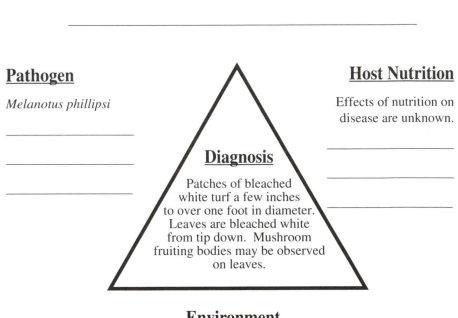

Melanotus phillipsi

Host Nutrition

Effects of nutrition on disease are unknown.

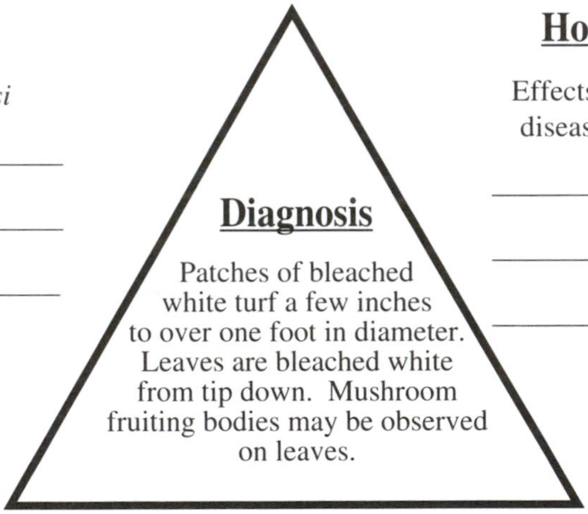

Diagnosis

Patches of bleached white turf a few inches to over one foot in diameter. Leaves are bleached white from tip down. Mushroom fruiting bodies may be observed on leaves.

Environment

Night temperatures > 70˚F (21˚C).
More than 10 hrs. of leaf wetness per day for several days.
Disease is particularly severe on soils from recently cleared forests.

White Patch

Disease Forecasting and Pathogen Detection

A white-patch forecaster and detection kits are not available.

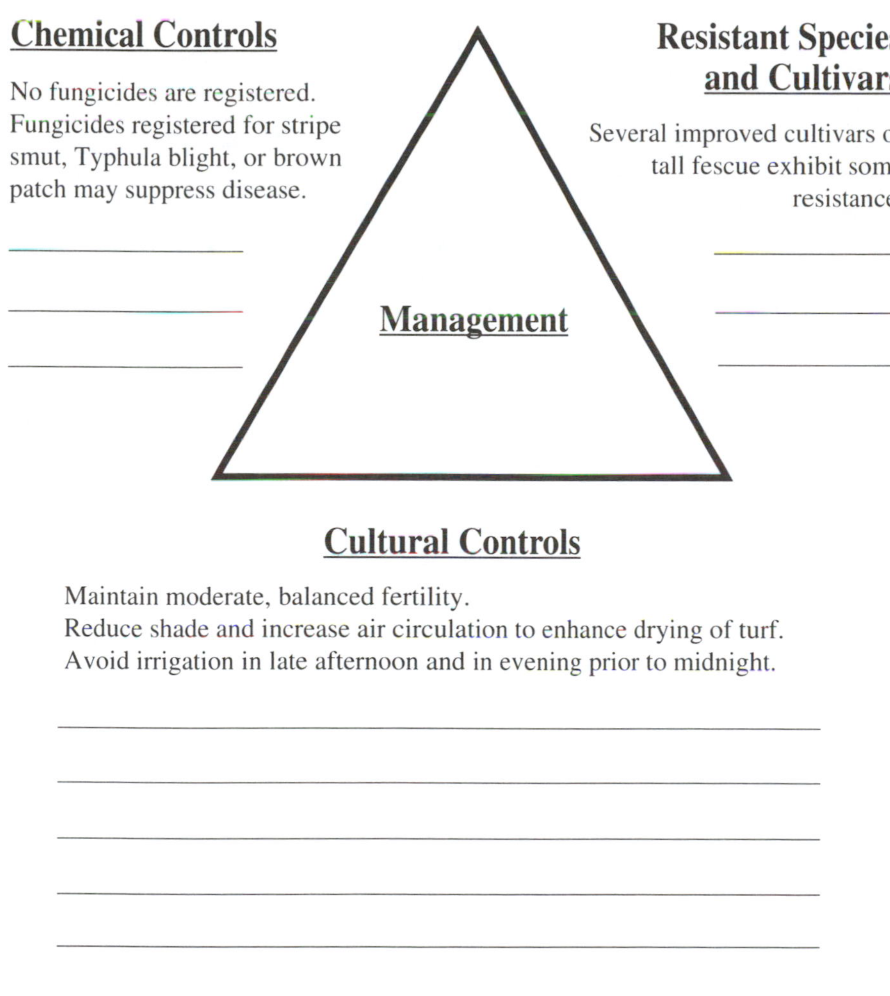

Chemical Controls

No fungicides are registered. Fungicides registered for stripe smut, Typhula blight, or brown patch may suppress disease.

Resistant Species and Cultivars

Several improved cultivars of tall fescue exhibit some resistance.

Cultural Controls

Maintain moderate, balanced fertility.
Reduce shade and increase air circulation to enhance drying of turf.
Avoid irrigation in late afternoon and in evening prior to midnight.

Yellow Patch

Hosts

Bluegrasses, bentgrasses and tall fescue.

Pathogen

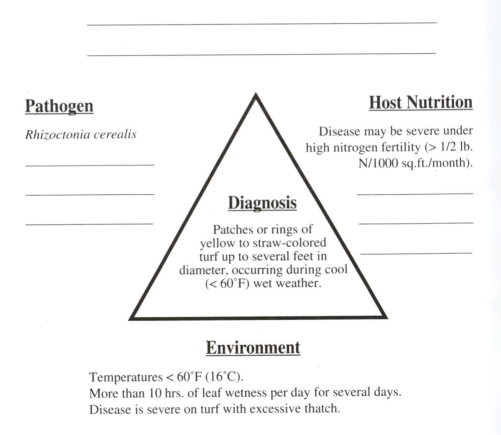

Rhizoctonia cerealis

Host Nutrition

Disease may be severe under high nitrogen fertility (> 1/2 lb. N/1000 sq.ft./month).

Diagnosis

Patches or rings of yellow to straw-colored turf up to several feet in diameter, occurring during cool (< 60°F) wet weather.

Environment

Temperatures < 60°F (16°C).
More than 10 hrs. of leaf wetness per day for several days.
Disease is severe on turf with excessive thatch.

Yellow Patch

Disease Forecasting and Pathogen Detection

A yellow patch forecaster and detection kits are not available.

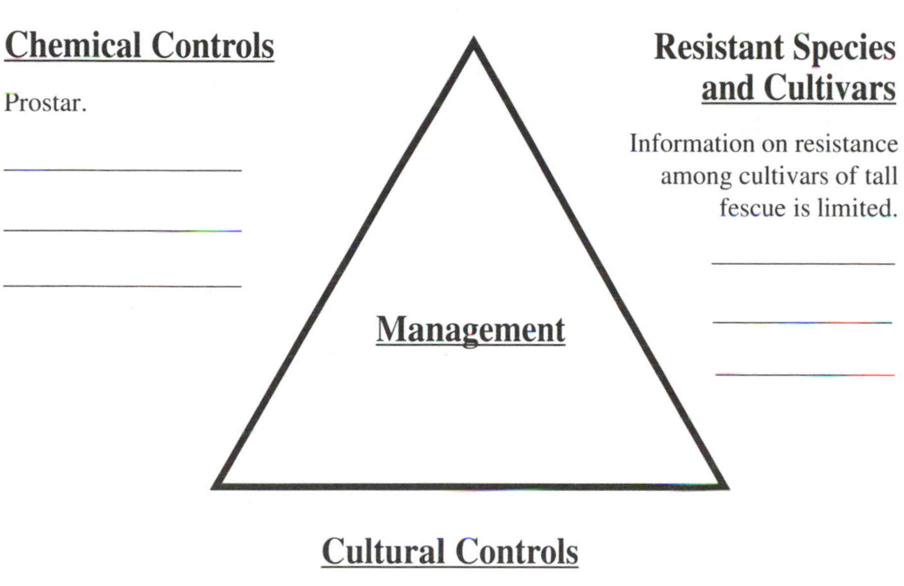

Chemical Controls

Prostar.

Resistant Species and Cultivars

Information on resistance among cultivars of tall fescue is limited.

Cultural Controls

Maintain moderate nitrogen fertility (1/2 lb. N/1000 sq.ft./month).
Maintain moderate to high levels of potash according to soil tests.
Reduce shade and increase air circulation to enhance drying of turf.
Reduce thatch thickness to 1/2 inch or less.

Chapter 4

Perennial Ryegrass

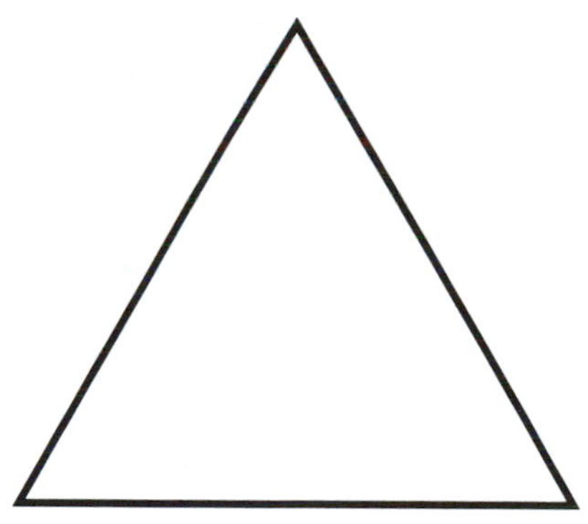

Anthracnose

Hosts

Creeping bentgrass, bluegrasses, fescues and perennial ryegrass. Annual bluegrass is particularly susceptible.

Pathogen

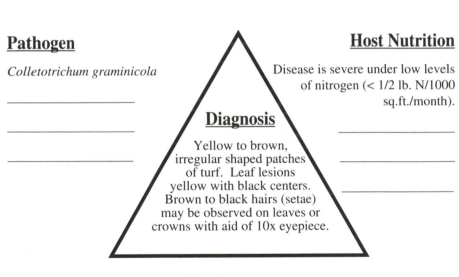

Colletotrichum graminicola

Host Nutrition

Disease is severe under low levels of nitrogen (< 1/2 lb. N/1000 sq.ft./month).

Diagnosis

Yellow to brown, irregular shaped patches of turf. Leaf lesions yellow with black centers. Brown to black hairs (setae) may be observed on leaves or crowns with aid of 10x eyepiece.

Environment

Temperature > 78°F (26°C).
More than 10 hrs. of leaf wetness per day for several days.
Disease is particularly severe on turf exposed to soil compaction and excess thatch.
(Pathogen may case crown rot of creeping bentgrass at temperatures from 60-77°F).

Anthracnose

Disease Forecasting and Pathogen Detection

An anthracnose forecaster is available from Neogen Corp., 620 Lesher Place, Lansing, MI 48912. Detection kits are not available.

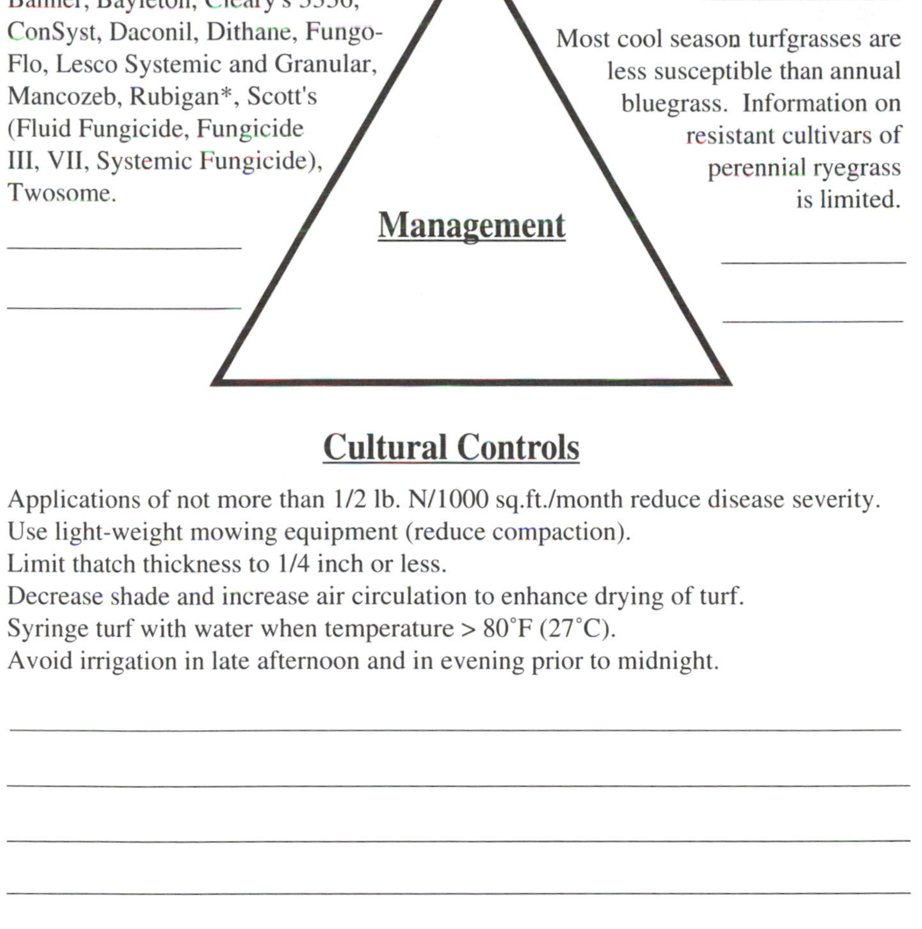

Chemical Controls

Banner, Bayleton, Cleary's 3336, ConSyst, Daconil, Dithane, Fungo-Flo, Lesco Systemic and Granular, Mancozeb, Rubigan*, Scott's (Fluid Fungicide, Fungicide III, VII, Systemic Fungicide), Twosome.

Resistant Species and Cultivars

Most cool season turfgrasses are less susceptible than annual bluegrass. Information on resistant cultivars of perennial ryegrass is limited.

Cultural Controls

Applications of not more than 1/2 lb. N/1000 sq.ft./month reduce disease severity.
Use light-weight mowing equipment (reduce compaction).
Limit thatch thickness to 1/4 inch or less.
Decrease shade and increase air circulation to enhance drying of turf.
Syringe turf with water when temperature > 80°F (27°C).
Avoid irrigation in late afternoon and in evening prior to midnight.

*May reduce populations of annual bluegrass.

Brown Blight

Hosts

Ryegrasses and fescues. Ryegrasses are particularly susceptible.

Pathogen

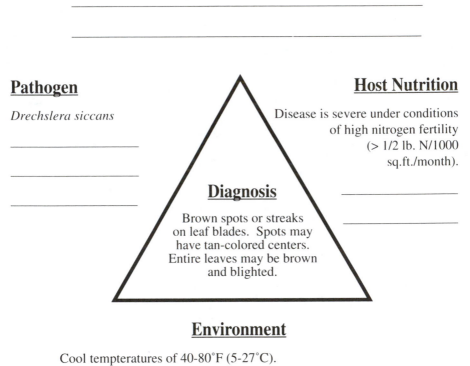

Drechslera siccans

Host Nutrition

Disease is severe under conditions of high nitrogen fertility (> 1/2 lb. N/1000 sq.ft./month).

Diagnosis

Brown spots or streaks on leaf blades. Spots may have tan-colored centers. Entire leaves may be brown and blighted.

Environment

Cool tempteratures of 40-80°F (5-27°C).
More than 10 hrs. of leaf wetness per day for several days.

Brown Blight

Disease Forecasting and Pathogen Detection

A brown blight forecaster and detection kits are not available.

Chemical Controls

Banner, Chipco 26019, ConSyst, Curalan, Daconil, Fore, Mancozeb, Terraclor*, Turfcide*, Touché, Twosome, Vorlan.

Resistant Species and Cultivars

Resistant cultivars of perennial ryegrass include: Dasher, Delray, Derby, Diplomat, Manhattan II, Omega II and Palmer.

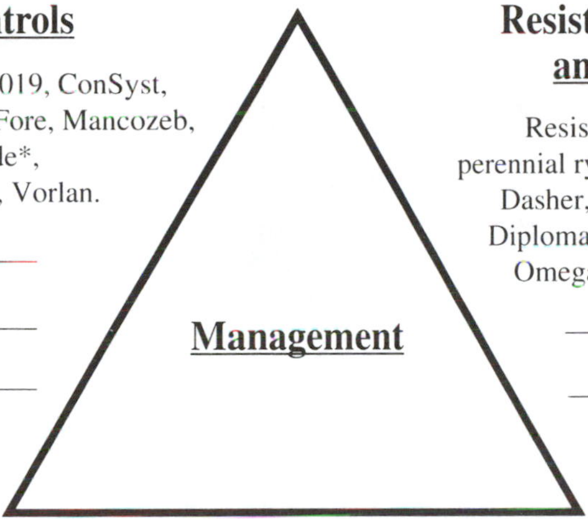

Management

Cultural Controls

Maintain moderate nitrogen fertility (1/2 lb. N/1000 sq.ft./month) and moderate to high levels of phosphorus and potash according to soil tests.
Decrease shade and increase air circulation to enhance drying of turf.
Avoid irrigation in late afternoon and in evening prior to midnight.
Raise mowing height.
Use light-weight mowing equipment to avoid stress on turf.

*Avoid application to bentgrasses.

Brown Patch
(Rhizoctonia Blight)

Hosts

All common species of turfgrasses.

Pathogen

Rhizoctonia solani

Diagnosis

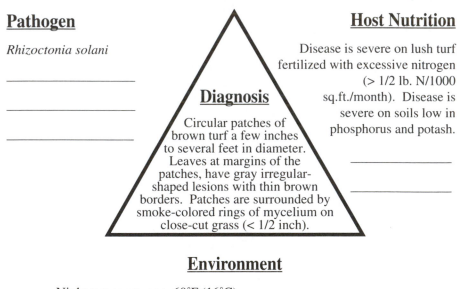

Circular patches of brown turf a few inches to several feet in diameter. Leaves at margins of the patches, have gray irregular-shaped lesions with thin brown borders. Patches are surrounded by smoke-colored rings of mycelium on close-cut grass (< 1/2 inch).

Host Nutrition

Disease is severe on lush turf fertilized with excessive nitrogen (> 1/2 lb. N/1000 sq.ft./month). Disease is severe on soils low in phosphorus and potash.

Environment

Night temperatures > 60˚F (16˚C).
More than 10 hrs. of foliar wetness per day for several days.
Disease is severe at low mowing heights (< 2 inches).

Brown Patch

Disease Forecasting and Pathogen Detection

A brown patch forecaster and detection kits are available from Neogen Corp., 620 Lesher Pl., Lansing, MI 48912.

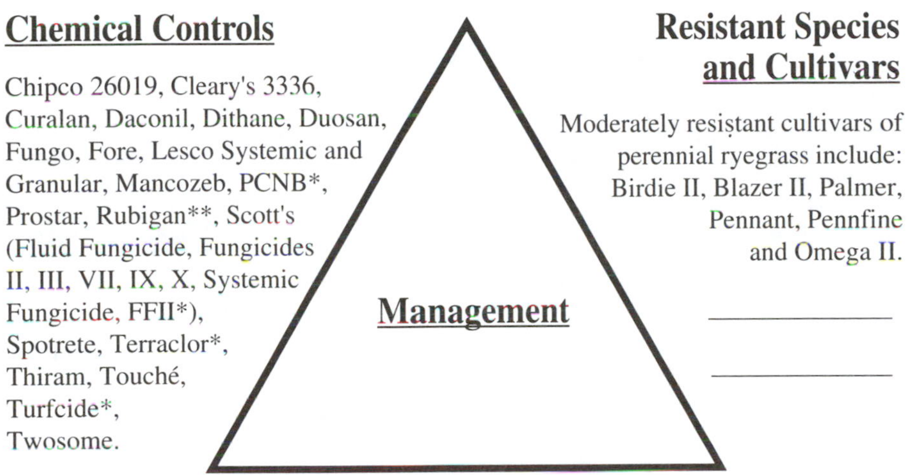

Chemical Controls

Chipco 26019, Cleary's 3336, Curalan, Daconil, Dithane, Duosan, Fungo, Fore, Lesco Systemic and Granular, Mancozeb, PCNB*, Prostar, Rubigan**, Scott's (Fluid Fungicide, Fungicides II, III, VII, IX, X, Systemic Fungicide, FFII*), Spotrete, Terraclor*, Thiram, Touché, Turfcide*, Twosome.

Resistant Species and Cultivars

Moderately resistant cultivars of perennial ryegrass include: Birdie II, Blazer II, Palmer, Pennant, Pennfine and Omega II.

Cultural Controls

Maintain moderate nitrogen fertility (1/2 lb. N/1000 sq.ft./month) and moderate to high levels of phosphorus and potash according to soil tests.
Decrease shade and increase air circulation to enhance drying of turf.
Avoid irrigation in late afternoon and in evening prior to midnight.
Maintain thatch at 1/4 inch thick or less.
Raise mowing height if possible.

*Avoid application to bentrasses. **May reduce populations of annual bluegrass.

Damping-off and Seed Rot

Hosts

All species of turfgrasses.

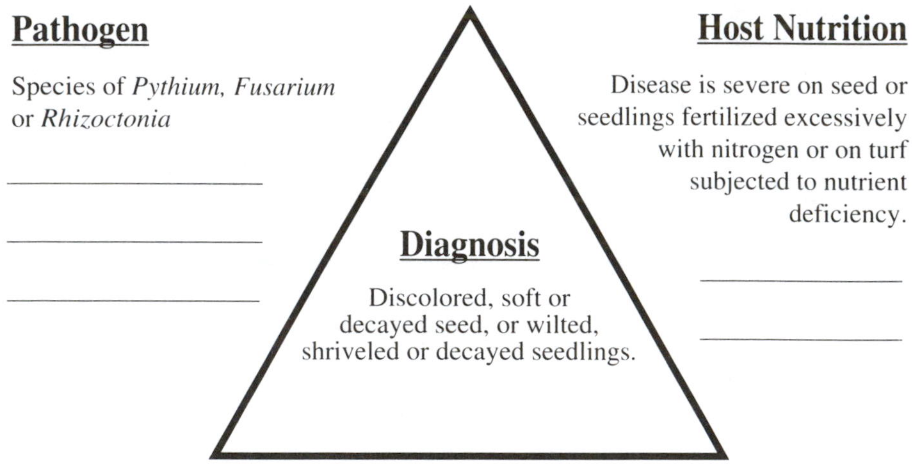

Pathogen

Species of *Pythium, Fusarium* or *Rhizoctonia*

Host Nutrition

Disease is severe on seed or seedlings fertilized excessively with nitrogen or on turf subjected to nutrient deficiency.

Diagnosis

Discolored, soft or decayed seed, or wilted, shriveled or decayed seedlings.

Environment

Temperatures too high (> 85˚F, 30˚C) or too low (< 60˚F, 15˚C) for optimum seedling development.
More than 10 hrs. of seed or seedling wetness per day for several days.
Excessive shade or overcrowding of seedlings.
Poor surface or subsurface drainage.

Damping-off and Seed Rot

Disease Forecasting and Pathogen Detection

A damping-off or seed rot forecaster and detection kits are not available.

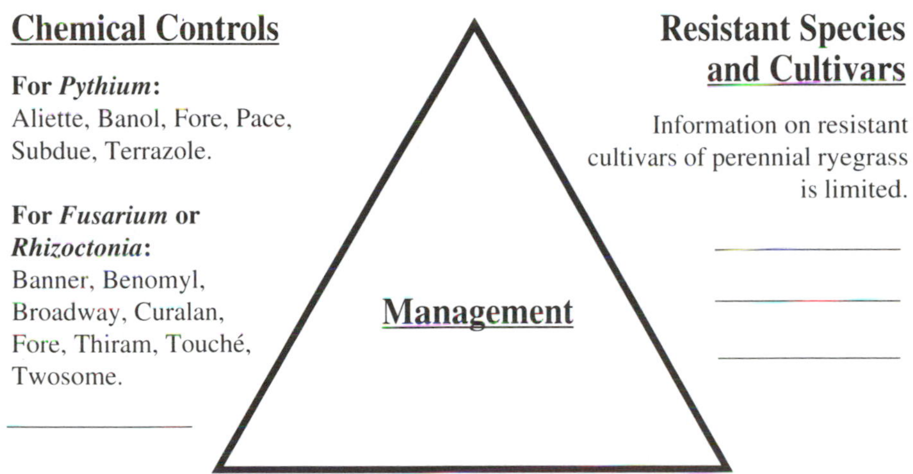

Chemical Controls

For *Pythium*:
Aliette, Banol, Fore, Pace, Subdue, Terrazole.

For *Fusarium* or *Rhizoctonia*:
Banner, Benomyl, Broadway, Curalan, Fore, Thiram, Touché, Twosome.

Resistant Species and Cultivars

Information on resistant cultivars of perennial ryegrass is limited.

Cultural Controls

Incorporate 1-3 lbs. N/1000 sq.ft. in seed bed prior to seeding.
Incorporate phosphorous and potash according to soil tests.
Fertilize seedlings with 1/2 to 1 lb. N/1000 sq.ft./month.
Seed turf when day temperatures are between 60° and 80°F (15° and 27°C).
Use recommended seeding rate (7-9 lb. seed/1000 sq.ft.) for perennial ryegrass.
Avoid high seeding rates.
Avoid irrigation in late afternoon and in evening prior to midnight.
Reduce shade and increase air circulation to enhance drying of turf.
Raise mowing height and use light-weight equipment.
Improve surface and subsurface drainage.

*Some materials may be sold as seed-treatments as well as foliar sprays.

Dollar Spot

Hosts

All common species of turfgrasses.

Pathogen

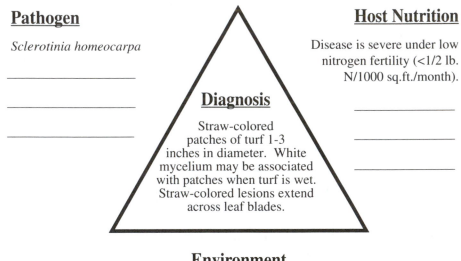

Sclerotinia homeocarpa

Host Nutrition

Disease is severe under low nitrogen fertility (<1/2 lb. N/1000 sq.ft./month).

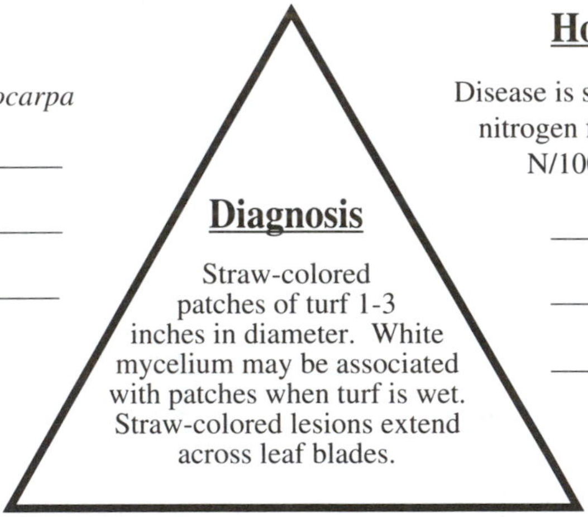

Diagnosis

Straw-colored patches of turf 1-3 inches in diameter. White mycelium may be associated with patches when turf is wet. Straw-colored lesions extend across leaf blades.

Environment

Night temperatures > 50°F (10°C) and day temperatures < 90°F (32°C).
More than 10 hrs. of leaf wetness per day for several days.
Disease is severe on turf subjected to drought stress.

Dollar Spot

Disease Forecasting and Pathogen Detection

A dollar spot forecaster is available from Pest Management Supply, P.O. Box 938, Amherst, MA 01004. Detection kits are available from Neogen Corp., 620 Lesher Pl., Lansing, MI 48912.

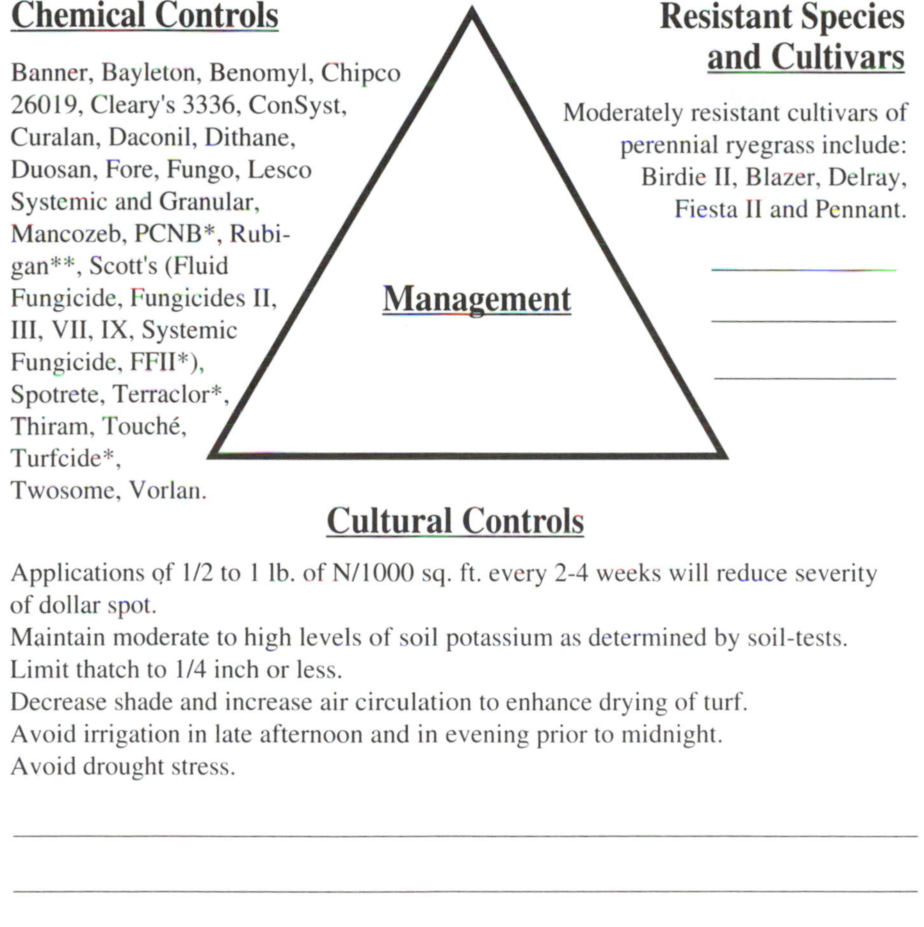

Chemical Controls

Banner, Bayleton, Benomyl, Chipco 26019, Cleary's 3336, ConSyst, Curalan, Daconil, Dithane, Duosan, Fore, Fungo, Lesco Systemic and Granular, Mancozeb, PCNB*, Rubigan**, Scott's (Fluid Fungicide, Fungicides II, III, VII, IX, Systemic Fungicide, FFII*), Spotrete, Terraclor*, Thiram, Touché, Turfcide*, Twosome, Vorlan.

Resistant Species and Cultivars

Moderately resistant cultivars of perennial ryegrass include: Birdie II, Blazer, Delray, Fiesta II and Pennant.

Cultural Controls

Applications of 1/2 to 1 lb. of N/1000 sq. ft. every 2-4 weeks will reduce severity of dollar spot.
Maintain moderate to high levels of soil potassium as determined by soil-tests.
Limit thatch to 1/4 inch or less.
Decrease shade and increase air circulation to enhance drying of turf.
Avoid irrigation in late afternoon and in evening prior to midnight.
Avoid drought stress.

*Avoid application to bentgrasses. **May reduce populations of annual bluegrass.

Fairy Ring

Hosts

All turfgrasses.

Pathogen

Several species of "mushroom-forming" fungi.

Host Nutrition

High nitrogen fertility (>1/2 lb. N/1000 sq.ft./month) may increase disease severity. Low nitrogen may increase the frequency of occurrence of fairy ring.

Diagnosis

Circles or archs of mushrooms or wilted, dead or dark green turf. White mats of fungal mycelium may be found in thatch or soil associated with circles or archs.

Environment

Light to moderate textured soils.
Soil pH of 5 to 7.5.
Low to moderate soil moisture.

Fairy Ring

Disease Forecasting and Pathogen Detection

A fairy ring forecaster and detection kits are not available.

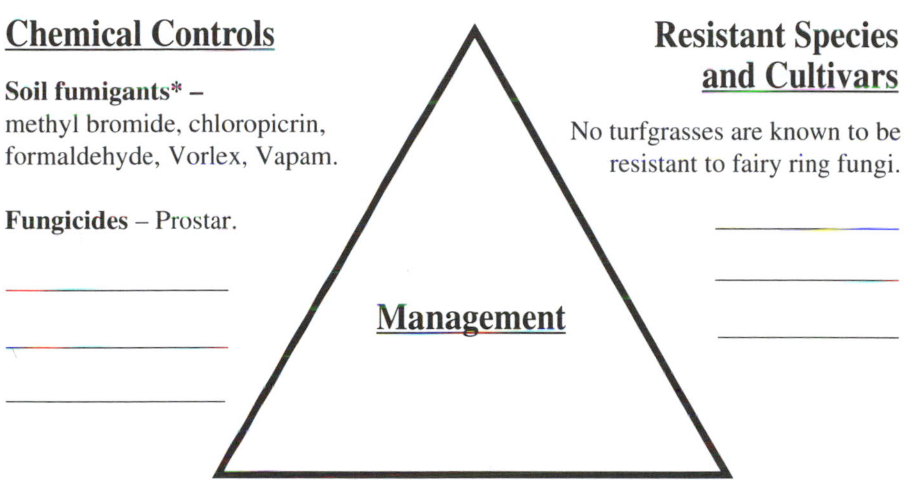

Chemical Controls

Soil fumigants* –
methyl bromide, chloropicrin, formaldehyde, Vorlex, Vapam.

Fungicides – Prostar.

Resistant Species and Cultivars

No turfgrasses are known to be resistant to fairy ring fungi.

Cultural Controls

Maintain moderate nitrogen fertility (1/2 lb. N/1000 sq.ft./month).
Maintain moderate to high levels of phosphorus and potash according to soil tests.
Excavate ring and soil 12 inches deep and 24 inches beyond ring or arch.
Replace with new soil.
Remove sod, cultivate soil 6 to 8 inches deep in several directions, add wetting agent to soil, reseed or sod.

* These chemicals are highly toxic to turfgrasses, animals and other life forms.

Fusarium Patch

Hosts

All cool season turfgrasses. Bentgrasses, annual bluegrass and perennial ryegrass are particularly susceptible.

Pathogen

Microdochium nivale

Host Nutrition

Disease is severe under high nitrogen fertility (>1/2 lb. N/1000 sq.ft./month) during late summer, fall, winter and spring. Disease may be severe on soils low in phosphorus and potash.

Diagnosis

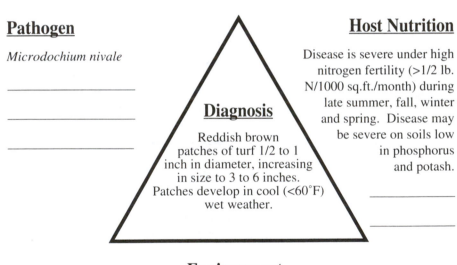

Reddish brown patches of turf 1/2 to 1 inch in diameter, increasing in size to 3 to 6 inches. Patches develop in cool (<60°F) wet weather.

Environment

Temperature < 60°F (16°C).
More than 10 hrs. of leaf wetness per day for several days.
Disease severity may be increased by applications of lime.

Fusarium Patch

Disease Forecasting and Pathogen Detection

A Fusarium patch forecaster and detection kits are not available.

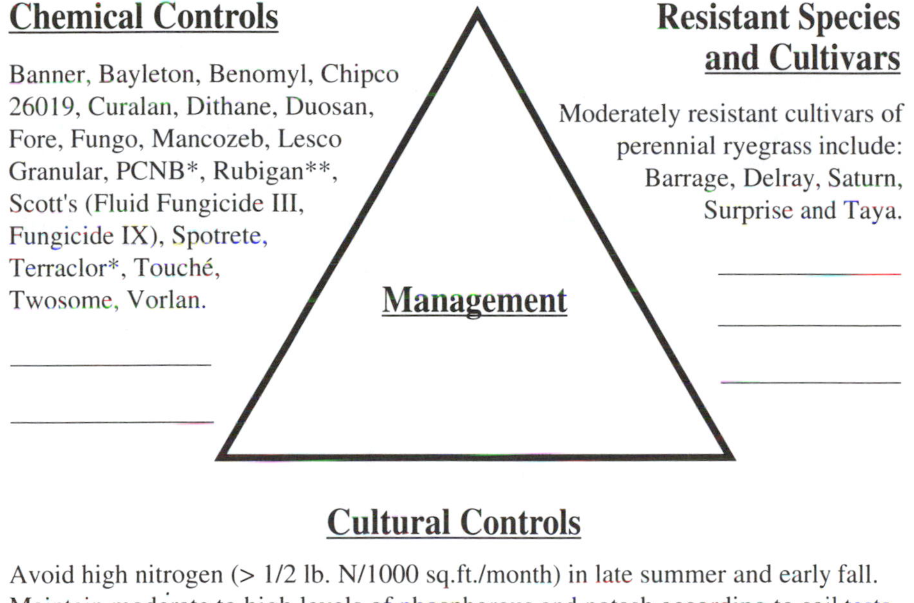

Chemical Controls

Banner, Bayleton, Benomyl, Chipco 26019, Curalan, Dithane, Duosan, Fore, Fungo, Mancozeb, Lesco Granular, PCNB*, Rubigan**, Scott's (Fluid Fungicide III, Fungicide IX), Spotrete, Terraclor*, Touché, Twosome, Vorlan.

Management

Resistant Species and Cultivars

Moderately resistant cultivars of perennial ryegrass include: Barrage, Delray, Saturn, Surprise and Taya.

Cultural Controls

Avoid high nitrogen (> 1/2 lb. N/1000 sq.ft./month) in late summer and early fall.
Maintain moderate to high levels of phosphorous and potash according to soil tests.
Decrease shade and increase air circulation to enhance drying of turf.
Avoid applications of lime if possible.
Avoid irrigation in late afternoon and in evening prior to midnight.

*Avoid application to actively growing bentgrass. **May reduce populations of annual bluegrass.

Gray Leaf Spot

Hosts

St. Augustine grass is particularly susceptible, but ryegrasses and fescues may exhibit severe symptoms under prolonged warm, wet conditions.

Pathogen

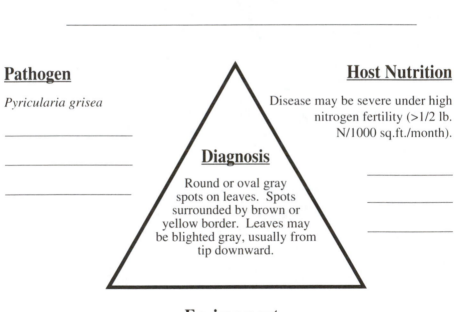

Pyricularia grisea

Host Nutrition

Disease may be severe under high nitrogen fertility (>1/2 lb. N/1000 sq.ft./month).

Diagnosis

Round or oval gray spots on leaves. Spots surrounded by brown or yellow border. Leaves may be blighted gray, usually from tip downward.

Environment

Night temperatures >70°F (21°C).
More than 10 hrs. of leaf wetness per day for several days.
Disease is severe in shaded areas or during periods of extended overcast weather.

Gray Leaf Spot

Disease Forecasting and Pathogen Detection

A gray leaf spot forecaster and detection kits are not available.

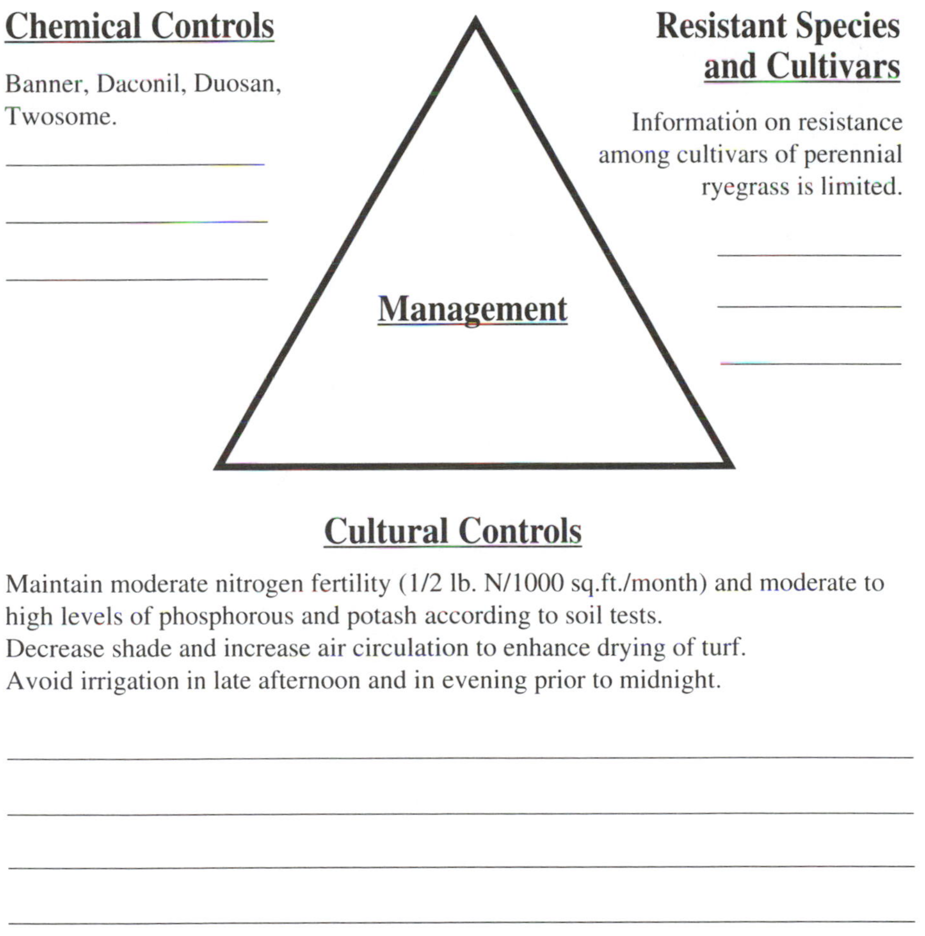

Chemical Controls

Banner, Daconil, Duosan, Twosome.

Resistant Species and Cultivars

Information on resistance among cultivars of perennial ryegrass is limited.

Cultural Controls

Maintain moderate nitrogen fertility (1/2 lb. N/1000 sq.ft./month) and moderate to high levels of phosphorous and potash according to soil tests.
Decrease shade and increase air circulation to enhance drying of turf.
Avoid irrigation in late afternoon and in evening prior to midnight.

Leaf Spot

Hosts

Bluegrasses, bentgrasses, fescues and perennial ryegrass.

Pathogen

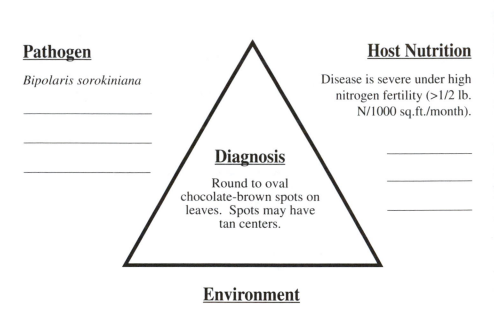

Bipolaris sorokiniana

Host Nutrition

Disease is severe under high nitrogen fertility (>1/2 lb. N/1000 sq.ft./month).

Diagnosis

Round to oval chocolate-brown spots on leaves. Spots may have tan centers.

Environment

Temperatures of 77°-95°F (25°-35°C).
Disease severity increases with increases in temperature.
More than 10 hrs. of leaf wetness per day for several days.

Leaf Spot

Disease Forecasting and Pathogen Detection

A leaf spot forecaster and detection kits are not available.

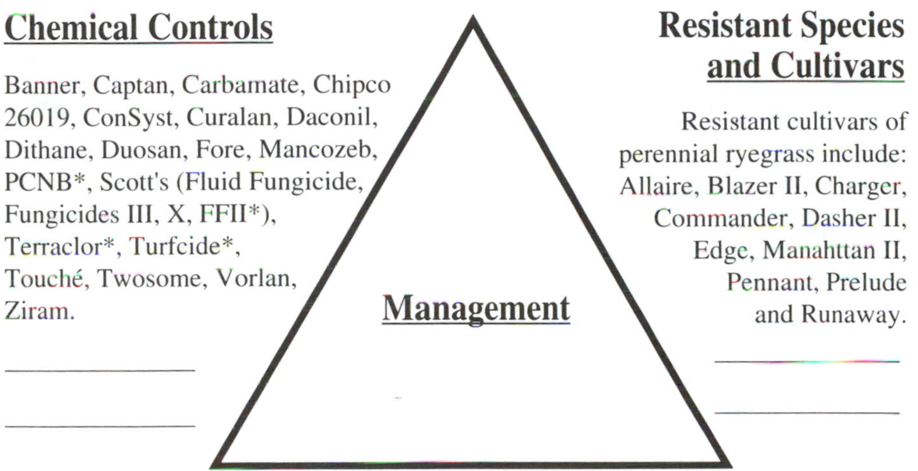

Chemical Controls

Banner, Captan, Carbamate, Chipco 26019, ConSyst, Curalan, Daconil, Dithane, Duosan, Fore, Mancozeb, PCNB*, Scott's (Fluid Fungicide, Fungicides III, X, FFII*), Terraclor*, Turfcide*, Touché, Twosome, Vorlan, Ziram.

Management

Resistant Species and Cultivars

Resistant cultivars of perennial ryegrass include: Allaire, Blazer II, Charger, Commander, Dasher II, Edge, Manahttan II, Pennant, Prelude and Runaway.

Cultural Controls

Apply moderate amounts of nitrogen during summer (1/4-1/2 lb. N/1000 sq.ft./ month).
Maintain moderate to high levels of soil P and K.
Decrease shade and increase air circulation to enhance drying of turf.
Avoid irrigation in late afternoon and in evening prior to midnight.
Limit thatch to 1/4 inch or less.
Raise mowing height.
Use light-weight mowing equipment to reduce stress.

*Avoid application to bentgrasses.

Perennial Ryegrass

Melting-Out

Hosts

Bluegrasses, ryegrasses and tall fescue.

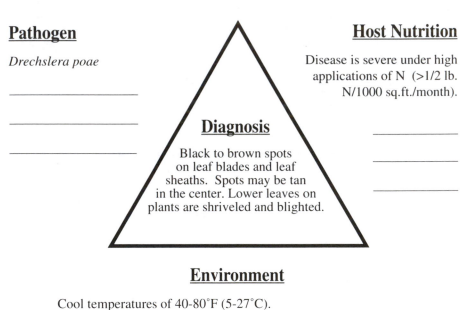

Pathogen

Drechslera poae

Host Nutrition

Disease is severe under high applications of N (>1/2 lb. N/1000 sq.ft./month).

Diagnosis

Black to brown spots on leaf blades and leaf sheaths. Spots may be tan in the center. Lower leaves on plants are shriveled and blighted.

Environment

Cool temperatures of 40-80°F (5-27°C).
More than 10 hrs. of leaf wetness per day for several days.
Mowing height < 2 inches.

Melting-Out

Disease Forecasting and Pathogen Detection

A melting-out forecaster and detection kits are not available.

Chemical Controls

Banner, Captan, Chipco 26019, ConSyst, Curalan, Daconil, Dithane, Fore, Mancozeb, Scott's (Fluid Fungicide, Fungicide III), Terraclor*, Turfcide*, Touché, Twosome, Vorlan.

Resistant Species and Cultivars

Information on resistance among cultivars of perennial ryegrass is limited.

Management

Cultural Controls

Fertilize with low to moderate levels of N during spring, summer and fall (1/4-1/2 lb. N/1000 sq.ft./month).
Decrease shade and increase air circulation to enhance drying of turf.
Avoid irrigation in late afternoon and in evening prior to midnight.
Raise mowing height.
Use light-weight mowing equipment to avoid stress on turf.
Limit thatch thickness to 1/4 inch or less.

*Avoid applications to bentgrasses.

Nematodes

Hosts

All common species of turfgrasses.

Pathogen

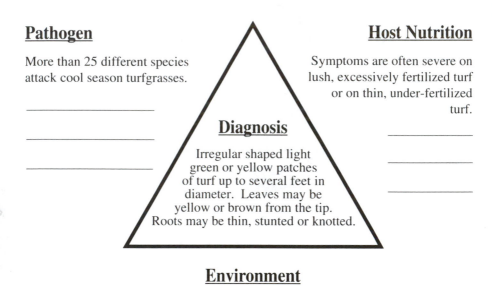

More than 25 different species attack cool season turfgrasses.

Host Nutrition

Symptoms are often severe on lush, excessively fertilized turf or on thin, under-fertilized turf.

Diagnosis

Irregular shaped light green or yellow patches of turf up to several feet in diameter. Leaves may be yellow or brown from the tip. Roots may be thin, stunted or knotted.

Environment

Soil temperatures > 40°F (5°C).
Symptoms are often severe on turf growing in sandy, light-textured soils.
Symptoms may be enhanced by drought and high temperatures (> 80°F, 26°C).

Nematodes

Disease Forecasting and Pathogen Detection

A nematode forecaster and detection kits are not available.

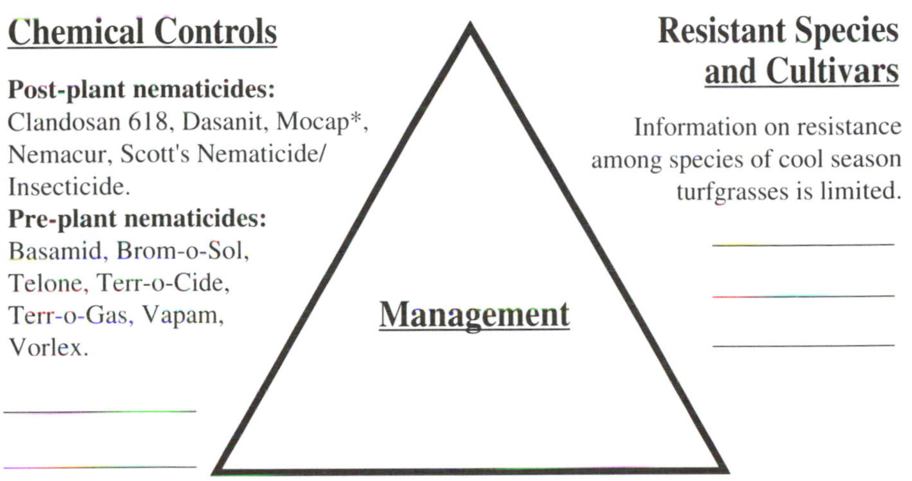

Chemical Controls

Post-plant nematicides: Clandosan 618, Dasanit, Mocap*, Nemacur, Scott's Nematicide/ Insecticide.
Pre-plant nematicides: Basamid, Brom-o-Sol, Telone, Terr-o-Cide, Terr-o-Gas, Vapam, Vorlex.

Resistant Species and Cultivars

Information on resistance among species of cool season turfgrasses is limited.

Cultural Controls

Maintain a balanced fertility program.
Apply 1/2 lb. N/1000 sq.ft./month during spring and summer.
Maintain moderate to high levels of phosphorus and potash according to soil tests.
Have soil analyzed for nematodes prior to seeding or sodding.
Use sod that is nematode-free.

*Avoid applications to bentgrasses.

Net Blotch

Hosts

Fescues, bluegrasses and ryegrasses.

Pathogen

Drechslera dictyoides

Host Nutrition

Disease is severe under high nitrogen fertility (> 1/2 lb. N/1000 sq.ft./month).

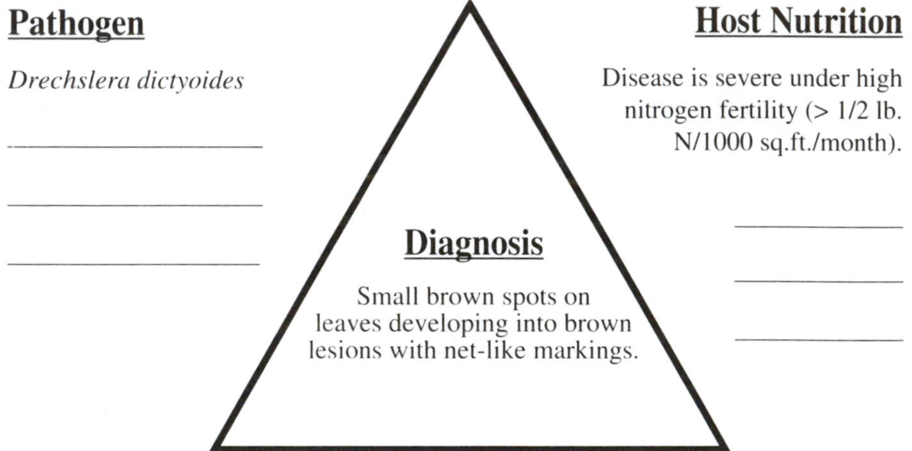

Diagnosis

Small brown spots on leaves developing into brown lesions with net-like markings.

Environment

Cool temperatures of 40-80°F (5-27°C).
More than 10 hrs. of leaf wetness per day for several days.
Mowing height < 2 inches.

Net Blotch

Disease Forecasting and Pathogen Detection

A net blotch forecaster and detection kits are not available.

Chemical Controls

Banner, Chipco 26019, ConSyst, Curalan, Daconil, Fore, Mancozeb, Terraclor*, Turfcide*, Touché, Twosome, Vorlan.

Resistant Species and Cultivars

Information on resistance among cultivars of perennial ryegrass is limited.

Management

Cultural Controls

Fertilize with low to moderate levels of N during spring, summer and fall (1/4-1/2 lb. N/1000 sq.ft./month).
Decrease shade and increase air circulation to enhance drying of turf.
Avoid irrigation in late afternoon and in evening prior to midnight.
Raise mowing height to at least 2 inches.
Use light-weight mowing equipment to avoid stress on turf.

*Avoid applications to bentgrasses.

Nigrospora Blight

Hosts

Kentucky bluegrass, fescues and perennial ryegrass.

Pathogen

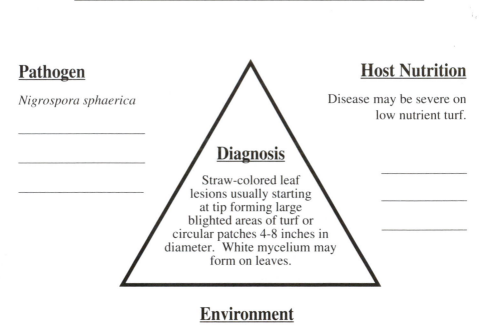

Nigrospora sphaerica

Host Nutrition

Disease may be severe on low nutrient turf.

Diagnosis

Straw-colored leaf lesions usually starting at tip forming large blighted areas of turf or circular patches 4-8 inches in diameter. White mycelium may form on leaves.

Environment

Temperature > 80°F (27°C).
More than 10 hrs. of leaf wetness per day for several days.
Disease may be severe on turf subjected to drought, herbicide stress or low mowing height.

Nigrospora Blight

Disease Forecasting and Pathogen Detection

A Nigrospora blight forecaster and detection kits are not available.

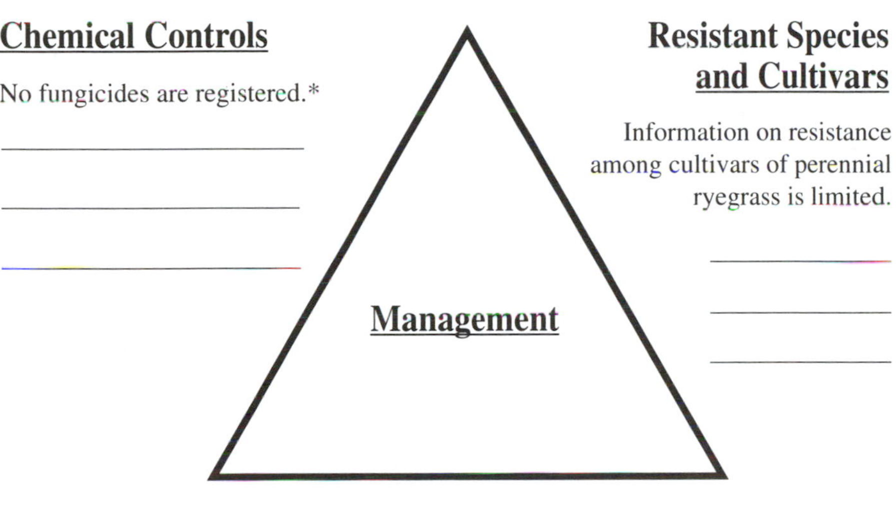

Chemical Controls

No fungicides are registered.*

Resistant Species and Cultivars

Information on resistance among cultivars of perennial ryegrass is limited.

Cultural Controls

Maintain moderate nitrogen fertility (1/2 lb. N/1000 sq.ft./month) during summer.
Maintain moderate to high levels of phosphorus and potash according to soil tests.
Avoid drought stress.
Avoid irrigation in late afternoon and evening prior to midnight.
Decrease shade and increase air circulation to enhance drying of turf.
Maintain turf at height of 2 inches or greater.
Avoid herbicide applications during summer.

* Chipco 26019 and Daconil 2787 provided control in experimental evaluations.

Pink Patch

Hosts

Perennial ryegrass, fine-leaf fescues, bentgrasses and bluegrasses.

Pathogen

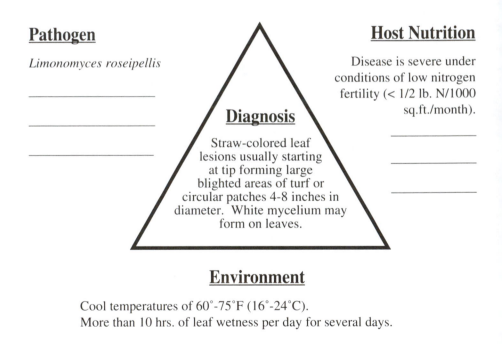

Limonomyces roseipellis

Host Nutrition

Disease is severe under conditions of low nitrogen fertility (< 1/2 lb. N/1000 sq.ft./month).

Diagnosis

Straw-colored leaf lesions usually starting at tip forming large blighted areas of turf or circular patches 4-8 inches in diameter. White mycelium may form on leaves.

Environment

Cool temperatures of 60°-75°F (16°-24°C).
More than 10 hrs. of leaf wetness per day for several days.

Pink Patch

Disease Forecasting and Pathogen Detection

A pink patch forecaster and detection kits are not available.

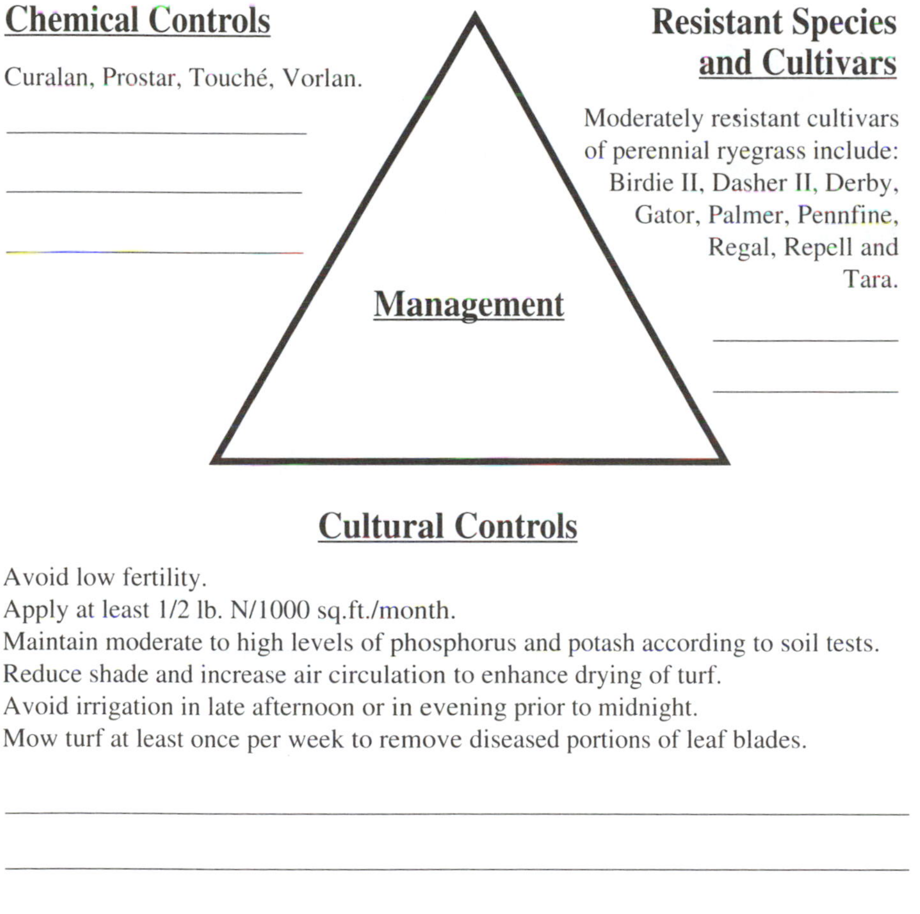

Chemical Controls
Curalan, Prostar, Touché, Vorlan.
Resistant Species and Cultivars
Moderately resistant cultivars of perennial ryegrass include: Birdie II, Dasher II, Derby, Gator, Palmer, Pennfine, Regal, Repell and Tara.
Management

Cultural Controls

Avoid low fertility.
Apply at least 1/2 lb. N/1000 sq.ft./month.
Maintain moderate to high levels of phosphorus and potash according to soil tests.
Reduce shade and increase air circulation to enhance drying of turf.
Avoid irrigation in late afternoon or in evening prior to midnight.
Mow turf at least once per week to remove diseased portions of leaf blades.

Pink Snow Mold

Hosts

All cool season grasses. Bentgrasses, annual bluegrass and perennial ryegrass are particularly susceptible.

Pathogen

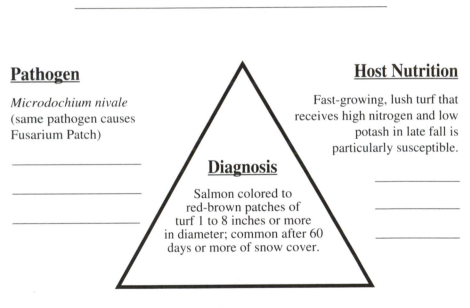

Microdochium nivale (same pathogen causes Fusarium Patch)

Host Nutrition

Fast-growing, lush turf that receives high nitrogen and low potash in late fall is particularly susceptible.

Diagnosis

Salmon colored to red-brown patches of turf 1 to 8 inches or more in diameter; common after 60 days or more of snow cover.

Environment

Disease is common after at least 60 days of snow cover, but pathogen can infect turf in absence of snow (see Fusarium Patch).
Disease is particularly severe when snow covers unfrozen ground.

Pink Snow Mold

Disease Forecasting and Pathogen Detection

A pink snow mold forecaster and detection kits are not available.

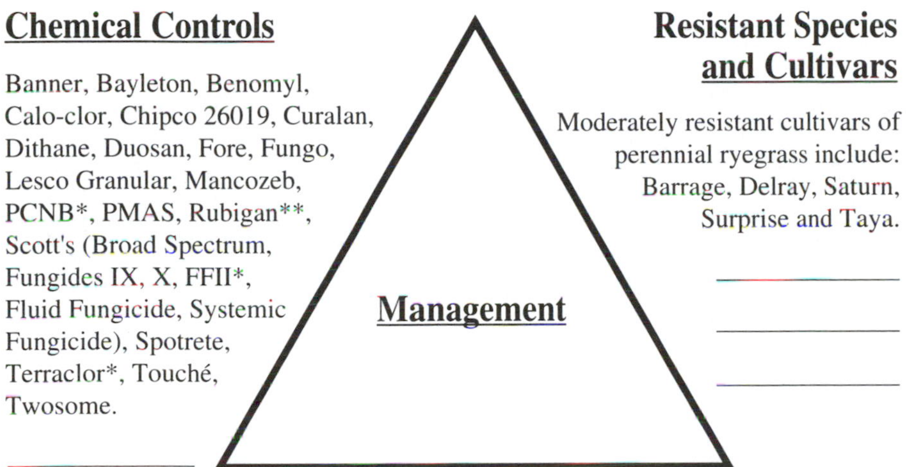

Chemical Controls

Banner, Bayleton, Benomyl, Calo-clor, Chipco 26019, Curalan, Dithane, Duosan, Fore, Fungo, Lesco Granular, Mancozeb, PCNB*, PMAS, Rubigan**, Scott's (Broad Spectrum, Fungides IX, X, FFII*, Fluid Fungicide, Systemic Fungicide), Spotrete, Terraclor*, Touché, Twosome.

Resistant Species and Cultivars

Moderately resistant cultivars of perennial ryegrass include: Barrage, Delray, Saturn, Surprise and Taya.

Cultural Controls

Maintain moderate nitrogen fertility (1/2 lb. N/1000 sq.ft./month) during late summer and fall.
Maintain high potash levels according to soil tests.
Use snow fence, shrubs or knolls as windbreaks to prevent excess snow from accumulating.
Prevent snow compaction by machinery or skiers.
Melt snow in spring with organic fertilizers.
Physically remove snow in spring.
Follow controls for Fusarium Patch after snow melt.

*Avoid applications to actively growing bentgrasses. **May reduce populations of annual bluegrass.

Pythium Blight

Hosts

All cool season grasses. Annual bluegrass and perennial ryegrass are particularly susceptible.

Pathogen

Pythium aphanidermatum and other species of *Pythium.*

Host Nutrition

Disease is severe under high nitrogen fertility (> 1/2 lb. N/1000 sq.ft./month). Deficiency in calcium may increase susceptibility.

Diagnosis

Greasy brown patches of turf an inch or less in diameter, increasing to several inches and turning straw colored. Grey-white cottony mycelium observed in early morning.

Environment

Night temperature > 65°F (18°C).
More than 10 hrs. of leaf wetness per day for several days.
Poor surface and sub-surface drainage.

Pythium Blight

Disease Forecasting and Pathogen Detection

Pythium blight forecasters are available from Pest Management Supply, P.O. Box 936, Amherst, MA 01004 or Neogen Corp., 620 Lesher Pl., Lansing, MI 48912. Detection kits are available from Neogen Corp.

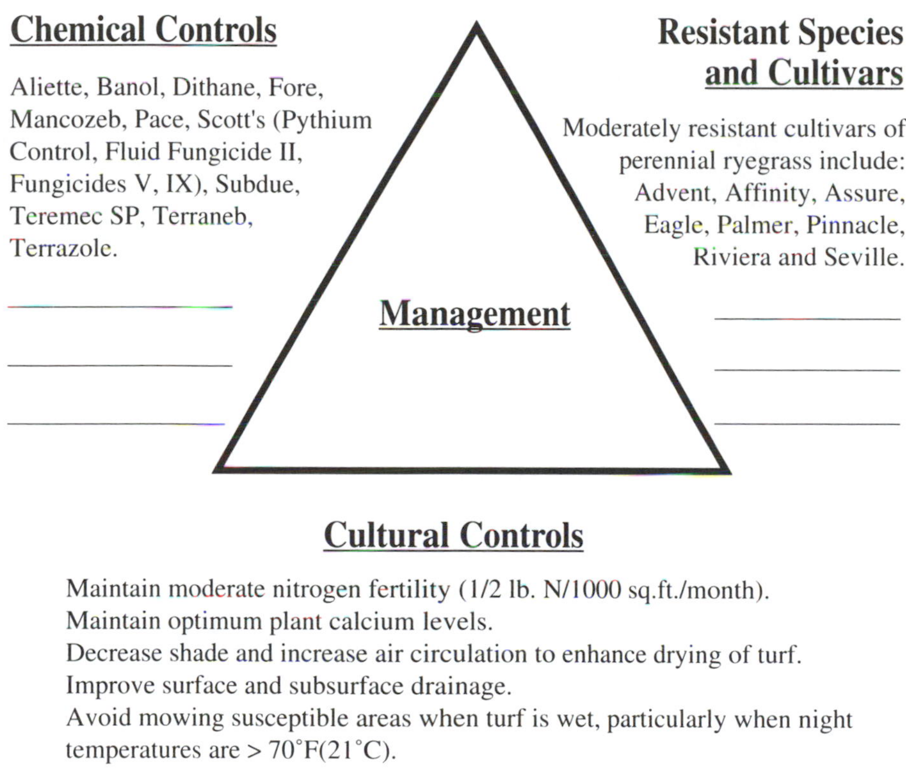

Chemical Controls

Aliette, Banol, Dithane, Fore, Mancozeb, Pace, Scott's (Pythium Control, Fluid Fungicide II, Fungicides V, IX), Subdue, Teremec SP, Terraneb, Terrazole.

Resistant Species and Cultivars

Moderately resistant cultivars of perennial ryegrass include: Advent, Affinity, Assure, Eagle, Palmer, Pinnacle, Riviera and Seville.

Cultural Controls

Maintain moderate nitrogen fertility (1/2 lb. N/1000 sq.ft./month).
Maintain optimum plant calcium levels.
Decrease shade and increase air circulation to enhance drying of turf.
Improve surface and subsurface drainage.
Avoid mowing susceptible areas when turf is wet, particularly when night temperatures are > 70°F(21°C).

Pythium Root Rot

Hosts

All species of cool season grasses.

Pathogen

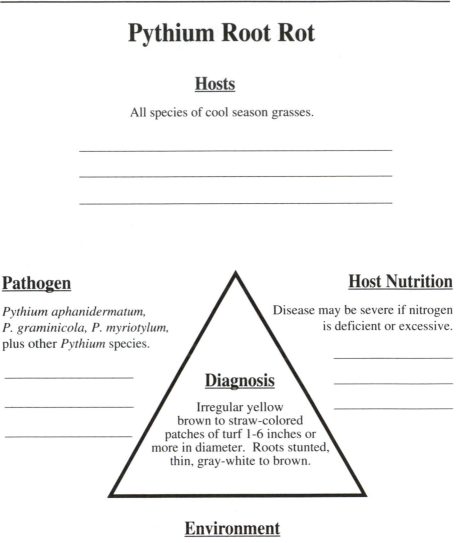

Pythium aphanidermatum, P. graminicola, P. myriotylum, plus other *Pythium* species.

Host Nutrition

Disease may be severe if nitrogen is deficient or excessive.

Diagnosis

Irregular yellow brown to straw-colored patches of turf 1-6 inches or more in diameter. Roots stunted, thin, gray-white to brown.

Environment

Cool (32°-50°F, 0-10°C) or warm (70°-90°F, 21-32°C) soil temperatures*.
High soil moisture.
Poor surface or subsurface drainage.
Conditions unfavorable for carbohydrate development by leaves–low light, low mowing height, excessive wear.

*Some *Pythium* species are favored by cool soils, other species by warm soils.

Pythium Root Rot

Disease Forecasting and Pathogen Detection

A Pythium root rot forecaster is not available. Detection kits are available from Neogen Corp., 620 Lesher Pl., Lansing, MI 48912.

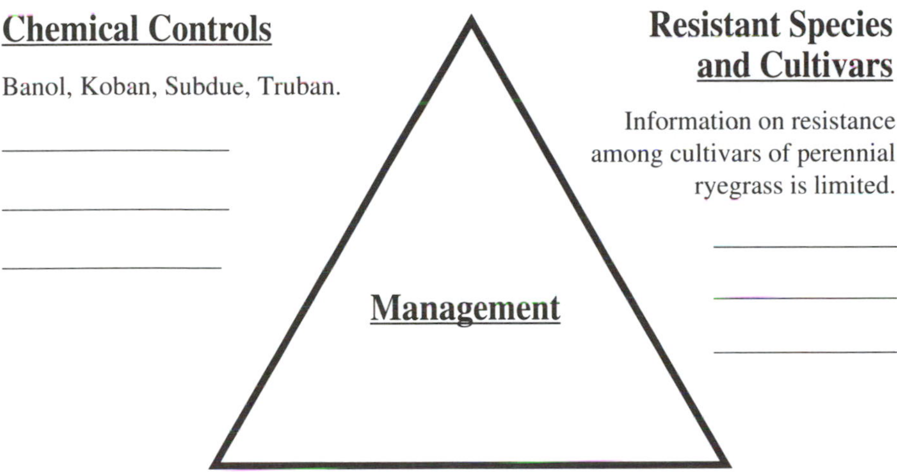

Chemical Controls

Banol, Koban, Subdue, Truban.

Resistant Species and Cultivars

Information on resistance among cultivars of perennial ryegrass is limited.

Cultural Controls

Maintain moderate levels of nitrogen (1/2 lb. N/1000 sq.ft./month).
Do not over fertilize with nitrogen in spring when roots are forming.
Maintain moderate to high levels of phosphorus and potash according to soil tests.
Improve surface and subsurface drainage.
Raise mowing height.
Decrease shade.
Use light-weight mowing equipment.
Applications of certain composts may reduce disease severity.

Red Thread

Hosts

Bentgrasses, bluegrasses, fine-leaf fescues and perennial ryegrass.
Fine-leaf fescues and perennial rye are particularly susceptible.

Pathogen

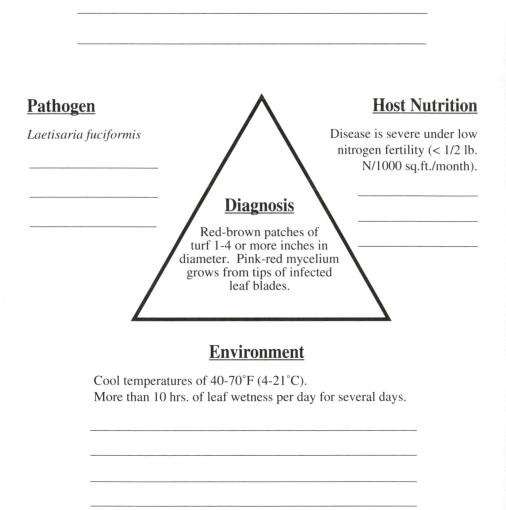

Laetisaria fuciformis

Host Nutrition

Disease is severe under low nitrogen fertility (< 1/2 lb. N/1000 sq.ft./month).

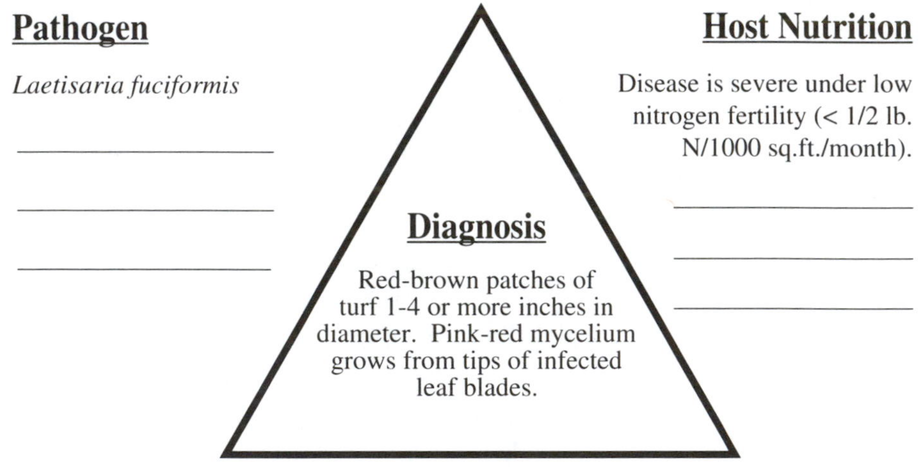

Diagnosis

Red-brown patches of turf 1-4 or more inches in diameter. Pink-red mycelium grows from tips of infected leaf blades.

Environment

Cool temperatures of 40-70°F (4-21°C).
More than 10 hrs. of leaf wetness per day for several days.

Red Thread

Disease Forecasting and Pathogen Detection

A red thread forecaster and detection kits are not available.

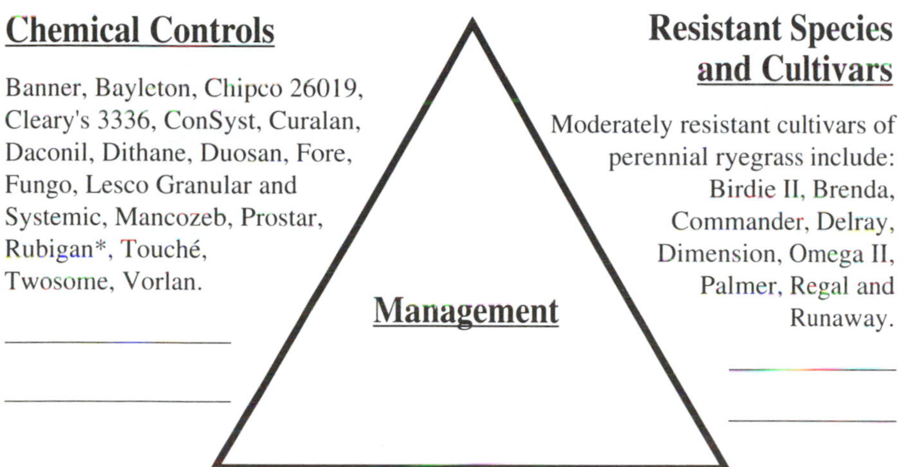

Chemical Controls

Banner, Bayleton, Chipco 26019, Cleary's 3336, ConSyst, Curalan, Daconil, Dithane, Duosan, Fore, Fungo, Lesco Granular and Systemic, Mancozeb, Prostar, Rubigan*, Touché, Twosome, Vorlan.

Management

Resistant Species and Cultivars

Moderately resistant cultivars of perennial ryegrass include: Birdie II, Brenda, Commander, Delray, Dimension, Omega II, Palmer, Regal and Runaway.

Cultural Controls

Fertilize with at least 1/2 lb. N/1000 sq.ft./month.
Maintain moderate to high levels of potash and phosphorus according to soil tests.
Reduce shade and increase air circulation to enhance drying of turf.
Avoid irrigation in late afternoon or in evening prior to midnight.
Maintain soil pH at 6.5 to 7.0.
Mow turf at least once per week to remove diseased portions of leaf blades.

*May reduce populations of annual bluegrass.

Rhizoctonia Leaf and Sheath Spot

Hosts

Bentgrasses, bluegrasses, perennial ryegrass and tall fescue.

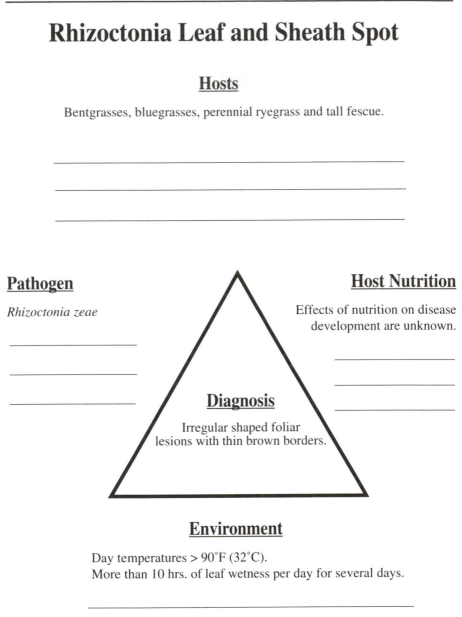

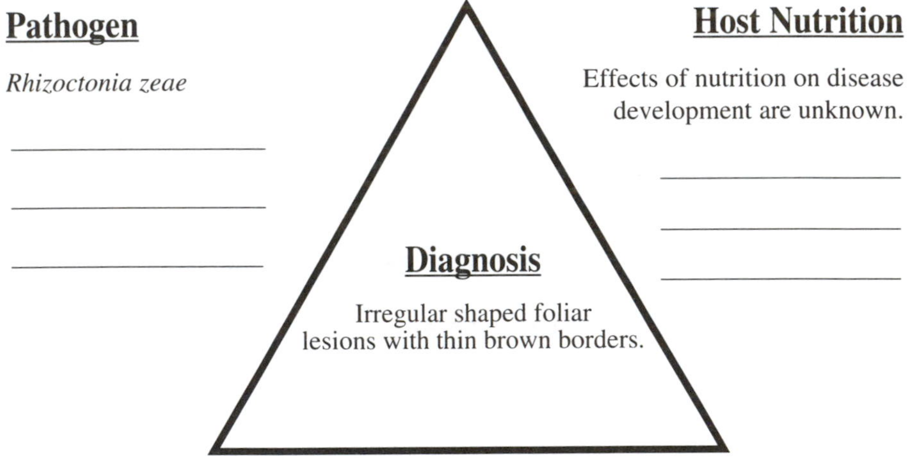

Pathogen

Rhizoctonia zeae

Host Nutrition

Effects of nutrition on disease development are unknown.

Diagnosis

Irregular shaped foliar lesions with thin brown borders.

Environment

Day temperatures > 90°F (32°C).
More than 10 hrs. of leaf wetness per day for several days.

Rhizoctonia Leaf and Sheath Spot

Disease Forecasting and Pathogen Detection

A leaf and sheath spot forecaster and detection kits are not available.

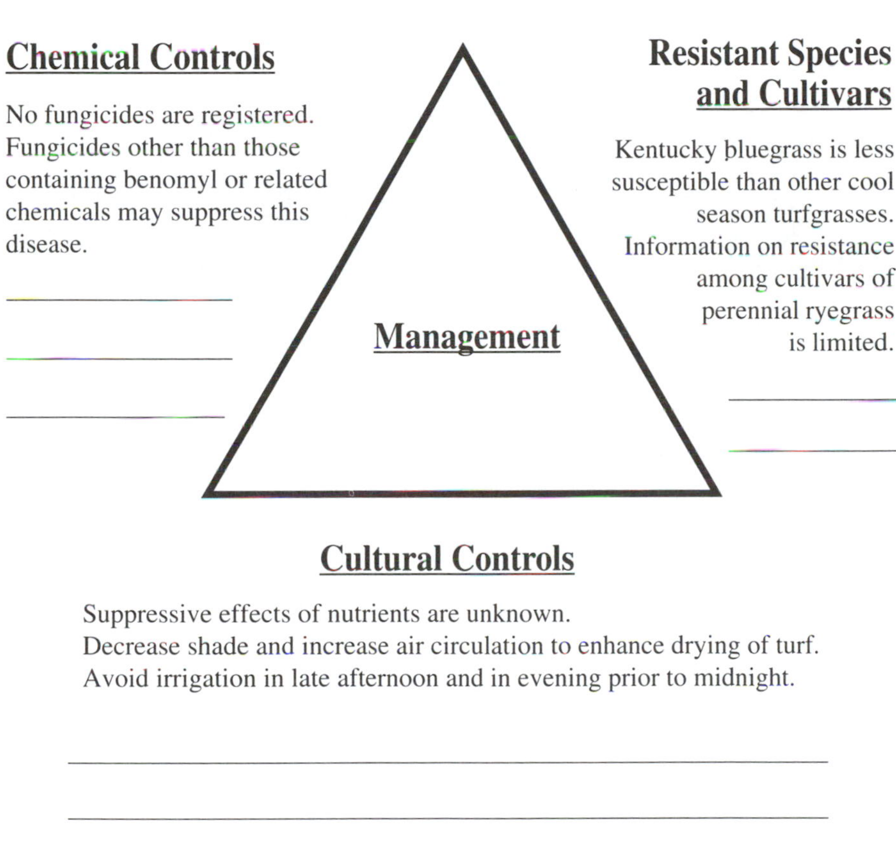

Chemical Controls

No fungicides are registered. Fungicides other than those containing benomyl or related chemicals may suppress this disease.

Resistant Species and Cultivars

Kentucky bluegrass is less susceptible than other cool season turfgrasses. Information on resistance among cultivars of perennial ryegrass is limited.

Cultural Controls

Suppressive effects of nutrients are unknown.
Decrease shade and increase air circulation to enhance drying of turf.
Avoid irrigation in late afternoon and in evening prior to midnight.

Rusts

Hosts

All common species of cool season turfgrasses.

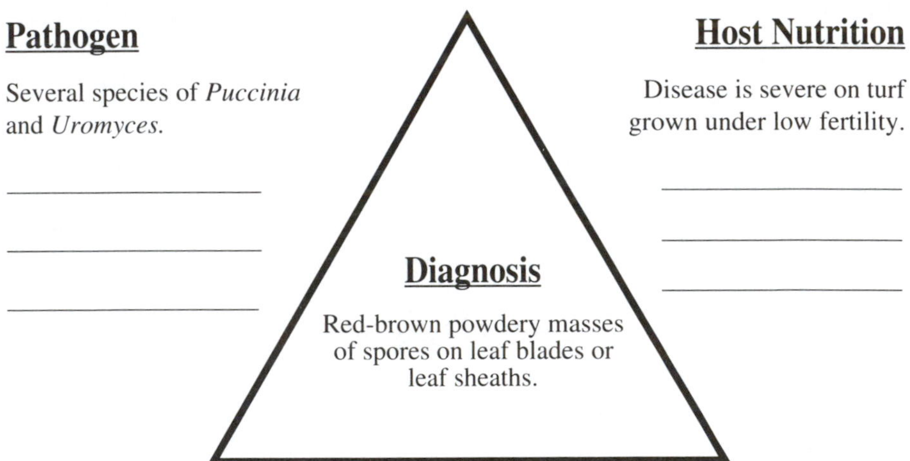

Pathogen

Several species of *Puccinia* and *Uromyces*.

Host Nutrition

Disease is severe on turf grown under low fertility.

Diagnosis

Red-brown powdery masses of spores on leaf blades or leaf sheaths.

Environment

Temperatures of 68°-86°F (20°-30°C).
Disease is severe on turf subjected to drought stress, low mowing, shade or poor air circulation.

Rust

Disease Forecasting and Pathogen Detection

A rust forecaster and detection kits are not available.

Chemical Controls

Banner, Bayleton, Captan, Carbamate, Cleary's 3336, ConSyst, Daconil, Dithane, Duosan, Fore, Mancozeb, Rubigan*, Scott's (Fluid Fungicide III, Fungicide VII, FFII**), Twosome, Ziram.

Resistant Species and Cultivars

Resistant cultivars of perennial ryegrass include: Barrage, Cowboy II, Cutless, Elite and Essence.

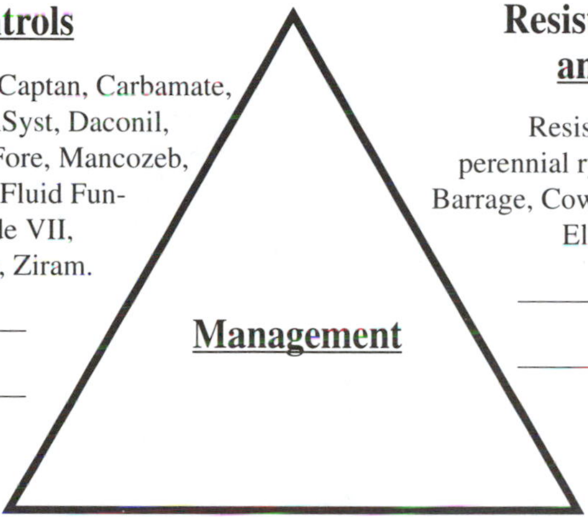

Management

Cultural Controls

Maintain moderate and balanced fertility throughout the growing season.
Reduce shade and increase air circulation to enhance drying of turf.
Increase mowing height.
Avoid drought stress.
Avoid irrigation in late afternoon and in evening prior to midnight.

*Avoid application to bentgrasses. **May reduce populations of annual bluegrass.

Stripe Smut

Hosts

Bluegrasses, bentgrasses, perennial rye and tall fescue. Kentucky bluegrass is more susceptible than other cool season grasses.

Pathogen

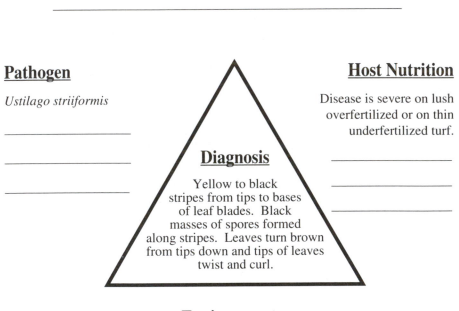

Ustilago striiformis

Host Nutrition

Disease is severe on lush overfertilized or on thin underfertilized turf.

Diagnosis

Yellow to black stripes from tips to bases of leaf blades. Black masses of spores formed along stripes. Leaves turn brown from tips down and tips of leaves twist and curl.

Environment

Infection occurs at 50°-68°F (10°-20°C).
Severe symptoms evident during drought and temperatures > 75°F (24°C).
Symptoms often more severe on acid soils and on turf with excessive thatch (> 1/2 inch thick).

Stripe Smut

Disease Forecasting and Pathogen Detection

A stripe smut forecaster and detection kits are not available.

Chemical Controls

Banner, Bayleton, Benomyl, Cleary's 3336, ConSyst, Fungo, Lesco Granular and Systemic, Scott's FFII*, Rubigan**.

Resistant Species and Cultivars

Information on resistance among cultivars of perennial ryegrass is limited.

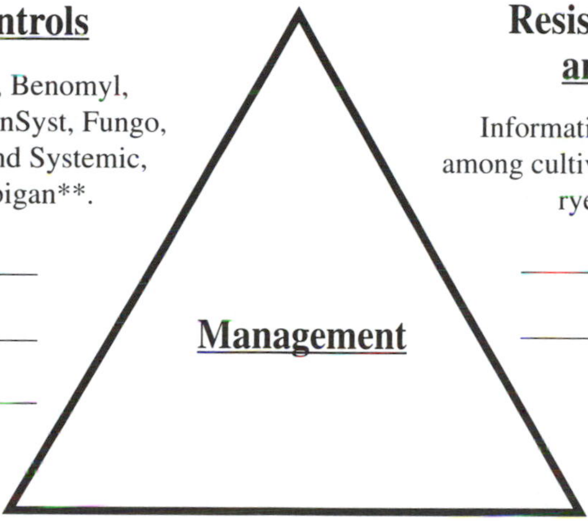

Management

Cultural Controls

Maintain moderate nitrogen fertility (1/2 lb. N/1000 sq.ft./month).
Maintain moderate phosphorus and high potash levels according to soil tests.
Avoid drought stress.
Apply lime if soil pH < 6.
Dethatch turf if thatch is > 1/2 inch thick.

*Avoid applications to bentgrasses. **May reduce populations of annual bluegrass.

Typhula Blight
(Gray Snow Mold)

Hosts

All cool season turfgrasses. Bentgrasses, annual bluegrass and perennial ryegrass are particularly susceptible.

Pathogen

Typhula incarnata and *T. ishikariensis*

Host Nutrition

Disease is severe on lush, fast-growing turf that is covered with snow. High nitrogen and low potash in fall can predispose turf to severe damage.

Diagnosis

Circular straw-colored patches of turf usually less than 10 inches in diameter, evident after snow-melt. Orange, brown to black sclerotia form on leaves.

Environment

Snow-cover is required for disease development. Disease is severe when snow-cover exceeds 90 days.

Typhula Blight

Disease Forecasting and Pathogen Detection

A Typhula blight forecaster and detection kits are not available.

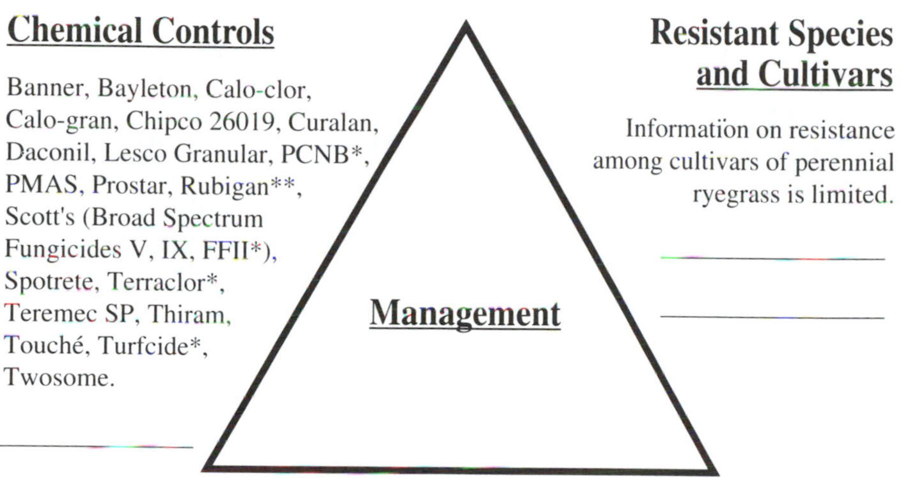

Chemical Controls

Banner, Bayleton, Calo-clor, Calo-gran, Chipco 26019, Curalan, Daconil, Lesco Granular, PCNB*, PMAS, Prostar, Rubigan**, Scott's (Broad Spectrum Fungicides V, IX, FFII*), Spotrete, Terraclor*, Teremec SP, Thiram, Touché, Turfcide*, Twosome.

Resistant Species and Cultivars

Information on resistance among cultivars of perennial ryegrass is limited.

Cultural Controls

Avoid a fertility program that results in lush, fast-growing turf in late fall and winter.
Maintain high potash levels according to soil tests.
Use snow fence, hedges or knolls to prevent snow form accumulating excessively on turf.
Use dark-colored organic fertilizers or composts to melt snow in spring.
Physically remove snow in spring.
Prevent compaction of snow during winter.

*Avoid application to actively growing bentgrasses. **May reduce populations of annual bluegrass.

Chapter 5

Fine-Leaf Fescues

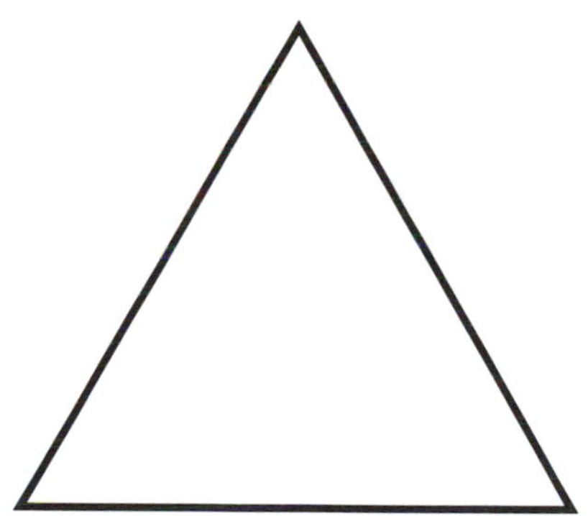

Anthracnose

Hosts

Creeping bentgrass, bluegrass, fescues, perennial ryegrass. Annual bluegrass is particularly susceptible.

Pathogen

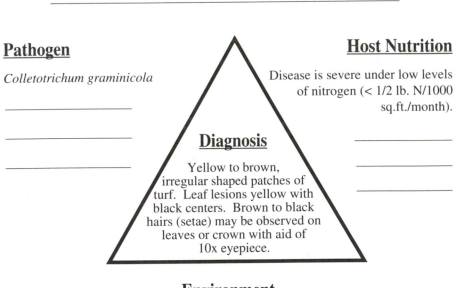

Colletotrichum graminicola

Host Nutrition

Disease is severe under low levels of nitrogen (< 1/2 lb. N/1000 sq.ft./month).

Diagnosis

Yellow to brown, irregular shaped patches of turf. Leaf lesions yellow with black centers. Brown to black hairs (setae) may be observed on leaves or crown with aid of 10x eyepiece.

Environment

Temperature > 78˚F (26˚C).
More than 10 hrs. of leaf wetness per day for several days.
Disease is particularly severe on turf exposed to soil compaction and excess thatch.
(Pathogen may cause crown rot of creeping bentgrass at temperatures from 60-77˚F).

Anthracnose

Disease Forecasting and Pathogen Detection

An anthracnose forecaster is available from Neogen Corp., 620 Lesher Pl., Lansing, MI 48912. Detection kits are not available.

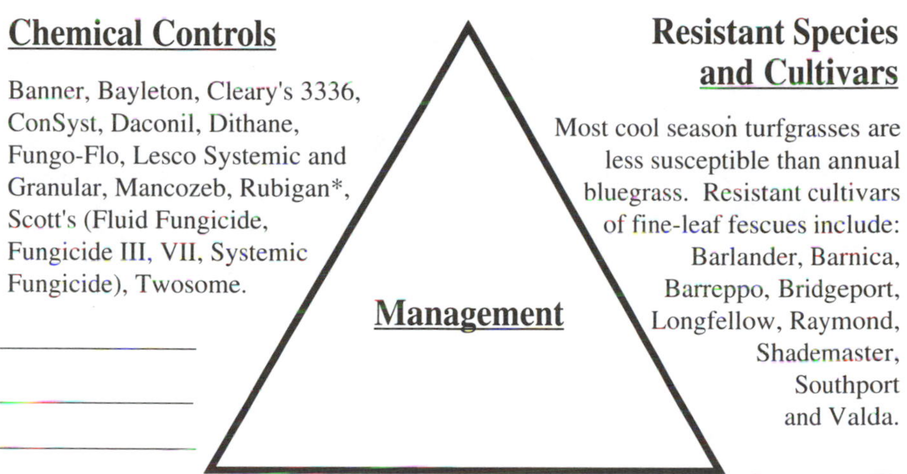

Chemical Controls

Banner, Bayleton, Cleary's 3336, ConSyst, Daconil, Dithane, Fungo-Flo, Lesco Systemic and Granular, Mancozeb, Rubigan*, Scott's (Fluid Fungicide, Fungicide III, VII, Systemic Fungicide), Twosome.

Resistant Species and Cultivars

Most cool season turfgrasses are less susceptible than annual bluegrass. Resistant cultivars of fine-leaf fescues include: Barlander, Barnica, Barreppo, Bridgeport, Longfellow, Raymond, Shademaster, Southport and Valda.

Cultural Controls

Applicaions of 1/2 lb. N/1000 sq.ft./month reduce disease severity.
Use light-weight mowing equipment (reduce compaction).
Limit thatch thickness to 1/4 inch or less.
Decrease shade and increase air circulation to enhance drying of turf.
Syringe turf with water when temperature > 80°F (27°C).
Avoid irrigation in late afternoon and in evening prior to midnight.

*May reduce populations of annual bluegrass.

Brown Patch
(Rhizoctonia Blight)

Hosts

All common species of turfgrasses.

Pathogen

Rhizoctonia solani

Host Nutrition

Disease is severe on lush turf fertilized with excessive nitrogen (> 1/2 lb. N/1000 sq.ft./month). Disease is severe on soils low in phosphorus and potash.

Diagnosis

Circular patches of brown turf a few inches to several feet in diameter. Leaves at margins of the patches have gray irregular-shaped lesions with thin brown borders.

Environment

Night temperatures > 60˚F (16˚C).
More than 10 hrs. of foliar wetness per day for several days.
Disease is severe at low mowing heights (< 2 inches).

Brown Patch

Disease Forecasting and Pathogen Detection

A brown patch forecaster and detection kits are available from Neogen Corp., 620 Lesher Pl., Lansing, MI 48912.

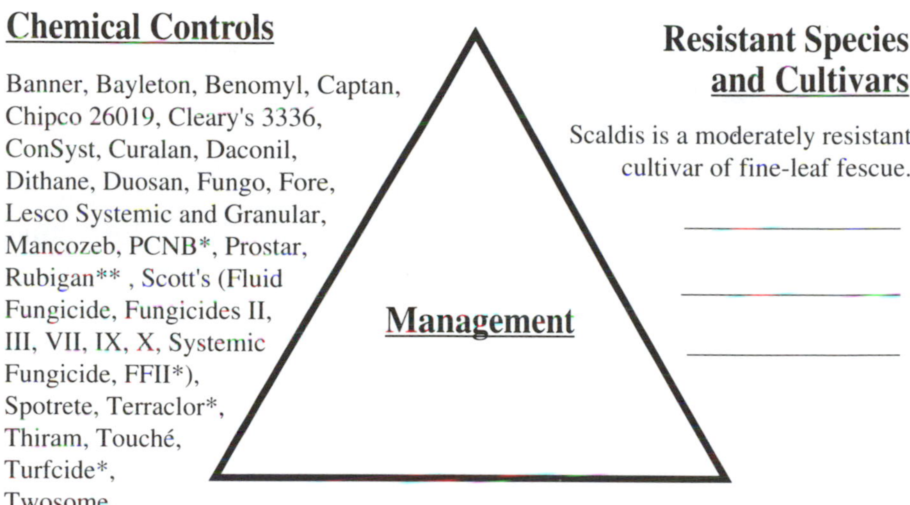

Chemical Controls

Banner, Bayleton, Benomyl, Captan, Chipco 26019, Cleary's 3336, ConSyst, Curalan, Daconil, Dithane, Duosan, Fungo, Fore, Lesco Systemic and Granular, Mancozeb, PCNB*, Prostar, Rubigan** , Scott's (Fluid Fungicide, Fungicides II, III, VII, IX, X, Systemic Fungicide, FFII*), Spotrete, Terraclor*, Thiram, Touché, Turfcide*, Twosome.

Resistant Species and Cultivars

Scaldis is a moderately resistant cultivar of fine-leaf fescue.

Cultural Controls

Maintain moderate nitrogen fertility (1/2 lb. N/1000 sq.ft./month).
Maintain moderate phosphorous and high potash according to soil tests.
Decrease shade and increase air circulation to enhance drying of turf.
Avoid irrigation in late afternoon and in evening prior to midnight.
Maintain thatch at 1/4 inch thick or less.
Raise mowing height if possible.

*Avoid application to bentgrasses. **May reduce populations of annual bluegrass.

Damping-off and Seed Rot

Hosts

All species of turfgrasses.

Pathogen

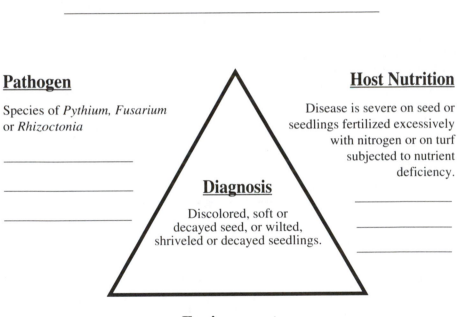

Species of *Pythium, Fusarium* or *Rhizoctonia*

Host Nutrition

Disease is severe on seed or seedlings fertilized excessively with nitrogen or on turf subjected to nutrient deficiency.

Diagnosis

Discolored, soft or decayed seed, or wilted, shriveled or decayed seedlings.

Environment

Temperatures too high (> 85°F, 30°C) or too low (< 60°F, 15°C) for optimum seedling development.
More than 10 hrs. of seed or seedling wetness per day for several days.
Excessive shade or overcrowding of seedlings.
Poor surface or subsurface drainage.

Damping-off and Seed Rot

Disease Forecasting and Pathogen Detection

A damping-off or seed rot forecaster is not available. Detection kits for ***Pythium*** and ***Rhizoctonia*** are available from Neogen Corp., 620 Lesher Pl., Lansing, MI 48912.

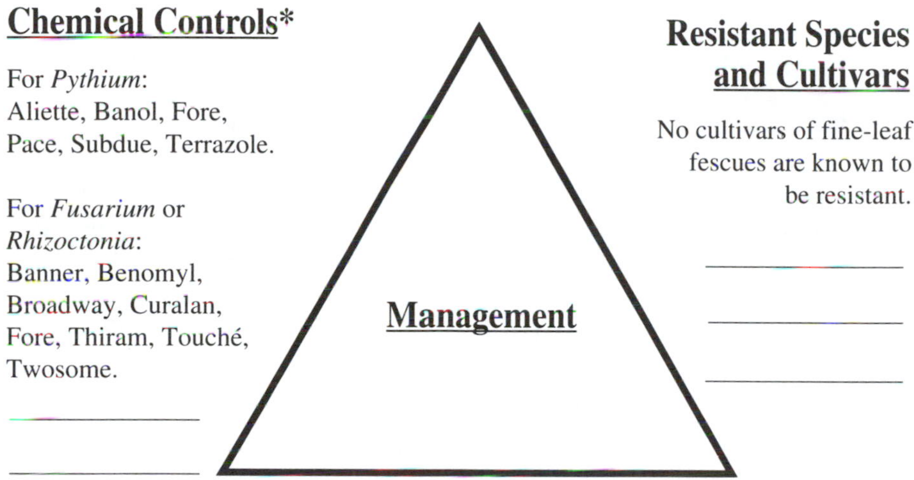

Chemical Controls*

For *Pythium*:
Aliette, Banol, Fore, Pace, Subdue, Terrazole.

For *Fusarium* or *Rhizoctonia*:
Banner, Benomyl, Broadway, Curalan, Fore, Thiram, Touché, Twosome.

Management

Resistant Species and Cultivars

No cultivars of fine-leaf fescues are known to be resistant.

Cultural Controls

Incorporate 1-3 lbs. N/1000 sq.ft. in seed bed prior to seeding.
Incorporate phosphorous and potash according to soil tests.
Fertilize seedlings with 1/2 to 1 lb. N/1000 sq.ft./month.
Seed turf when day temperatures are between 60° and 80°F (15° and 27°C).
Use recommended seeding rate (4 to 5 lb. seed/1000 sq.ft.) for fine-leaf fescues.
Avoid high seeding rates.
Avoid irrigation in late afternoon and in evening prior to midnight.
Reduce shade and increase air circulation to enhance drying of turf.
Raise mowing height and use light-weight equipment.
Improve surface and subsurface drainage.

* Some materials may be sold as seed-treatments as well as foliar sprays.

Dollar Spot

Hosts

All common species of turfgrasses.

Pathogen

Sclerotinia homeocarpa

Host Nutrition

Disease is severe under low nitrogen fertility (< 1/2 lb. N/1000 sq.ft./month).

Diagnosis

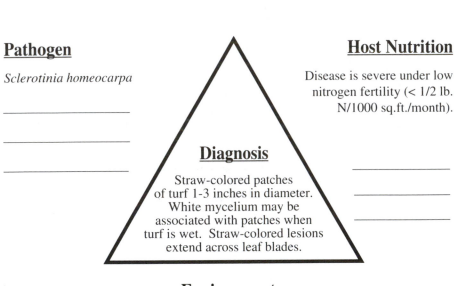

Straw-colored patches of turf 1-3 inches in diameter. White mycelium may be associated with patches when turf is wet. Straw-colored lesions extend across leaf blades.

Environment

Night temperatures > 50˚F (10˚C) and day temperatures < 90˚F (32˚C).
More than 10 hrs. of leaf wetness per day for several days.
Disease is severe on turf subjected to drought stress.

Dollar Spot

Disease Forecasting and Pathogen Detection

A dollar spot forecaster is available from
Pest Management Suply, P.O. Box 938, Amherst, MA 01004.
Detection kits are available from
Neogen Corp., 620 Lesher Pl., Lansing, MI 48912.

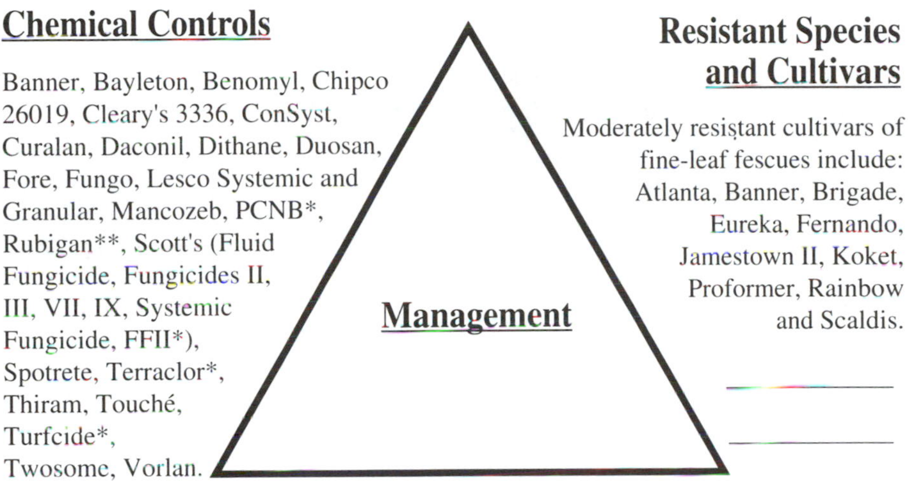

Chemical Controls

Banner, Bayleton, Benomyl, Chipco 26019, Cleary's 3336, ConSyst, Curalan, Daconil, Dithane, Duosan, Fore, Fungo, Lesco Systemic and Granular, Mancozeb, PCNB*, Rubigan**, Scott's (Fluid Fungicide, Fungicides II, III, VII, IX, Systemic Fungicide, FFII*), Spotrete, Terraclor*, Thiram, Touché, Turfcide*, Twosome, Vorlan.

Resistant Species and Cultivars

Moderately resistant cultivars of fine-leaf fescues include: Atlanta, Banner, Brigade, Eureka, Fernando, Jamestown II, Koket, Proformer, Rainbow and Scaldis.

Cultural Controls

Applications of 1/2 to 1 lb. of N/1000 sq.ft. every 2-4 weeks will reduce severity of dollar spot.
Maintain moderate to high levels of soil potassium as determined by soil tests.
Limit thatch to 1/4 inch or less.
Decrease shade and increase air circulation to enhance drying of turf.
Avoid irrigation in late afternoon and in evening prior to midnight.
Avoid drought stress.

*Avoid application to bentgrasses. **May reduce populations of annual bluegrass.

Fairy Ring

Hosts

All turfgrasses.

Pathogen

Several species of "mushroom-forming" fungi.

Host Nutrition

High nitrogen fertility (> 1/2 lb. N/1000 sq.ft./month) may increase disease severity. Low nitrogen may increase the frequency of occurrence of fairy ring.

Diagnosis

Circles or archs of mushrooms or wilted, dead or dark green turf. White mats of fungal mycelium may be found in thatch or soil associated with circles or archs.

Environment

Light to moderate textured soils.
Soil pH of 5 to 7.5.
Low to moderate soil moisture.

Fairy Ring

Disease Forecasting and Pathogen Detection

A fairy ring forecaster and detection kits are not available.

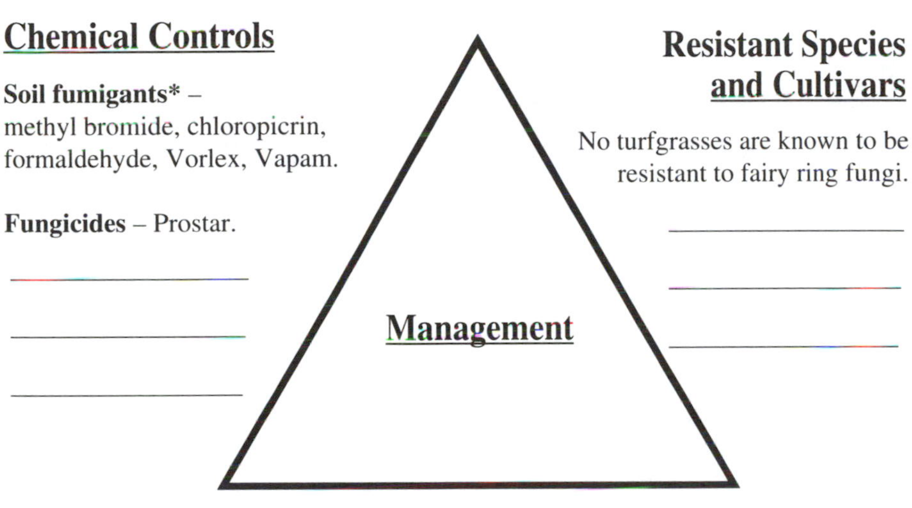

Chemical Controls

Soil fumigants* –
methyl bromide, chloropicrin, formaldehyde, Vorlex, Vapam.

Fungicides – Prostar.

Resistant Species and Cultivars

No turfgrasses are known to be resistant to fairy ring fungi.

Cultural Controls

Maintain moderate nitrogen fertility (1/2 lb. N/1000 sq.ft./month).
Maintain moderate to high levels of phosphorus and potash according to soil tests.
Excavate ring and soil 12 inches deep and 24 inches beyond ring or arch. Replace with new soil.
Remove sod, cultivate soil 6 ot 8 inches deep in several directions, add wetting agent to soil, reseed or sod.

* These chemicals are highly toxic to turfgrasses, animals and other life forms.

Fusarium Patch

Hosts

All cool season turfgrasses. Annual bluegrass, bentgrasses and perennial ryegrass are particularly susceptible.

Pathogen

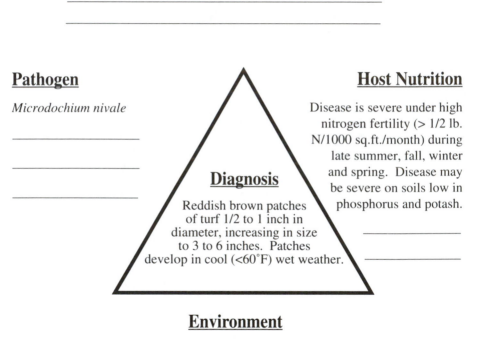

Microdochium nivale

Host Nutrition

Disease is severe under high nitrogen fertility (> 1/2 lb. N/1000 sq.ft./month) during late summer, fall, winter and spring. Disease may be severe on soils low in phosphorus and potash.

Diagnosis

Reddish brown patches of turf 1/2 to 1 inch in diameter, increasing in size to 3 to 6 inches. Patches develop in cool (<60°F) wet weather.

Environment

Temperature < 60°F (16°C).
More than 10 hrs. of leaf wetness per day for several days.
Disease severity may be increased by applications of lime.

Fusarium Patch

Disease Forecasting and Pathogen Detection

A Fusarium patch forecaster and detection kits are not available.

Chemical Controls

Banner, Bayleton, Benomyl, Chipco 26019, Curalan, Dithane, Duosan, Fore, Fungo, Mancozeb, Lesco Granular, PCNB*, Rubigan** , Scott's (Fluid Fungicide III, Fungicide IX), Spotrete, Terraclor*, Touché, Twosome, Vorlan.

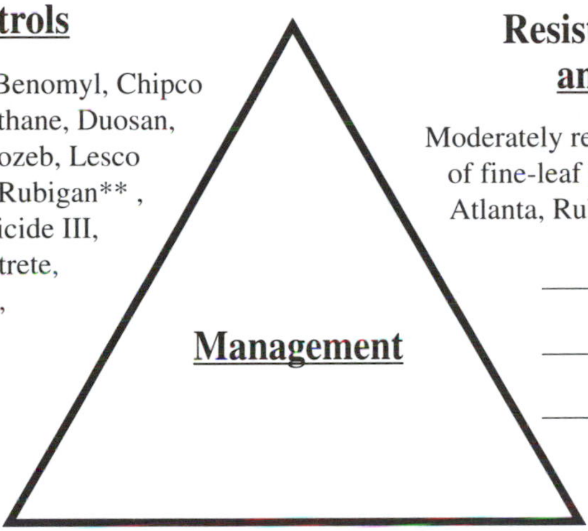

Management

Resistant Species and Cultivars

Moderately resistant cultivars of fine-leaf fescues include: Atlanta, Ruby and Dawson.

Cultural Controls

Avoid high nitrogen (> 1/2 lb. N/1000 sq.ft./month) in late summer and early fall.

Maintain moderate to high levels of phosphorous and potash according to soil tests.

Decrease shade and increase air circulation to enhance drying of turf.

Avoid applications of lime if possible.

Avoid irrigation in late afternoon and in evening prior to midnight.

*Avoid application to actively growing bentgrass. **May reduce populations of annual bluegrass.

Leaf Spot

Hosts

Bluegrasses, bentgrasses, fescues, and perennial ryegrass.

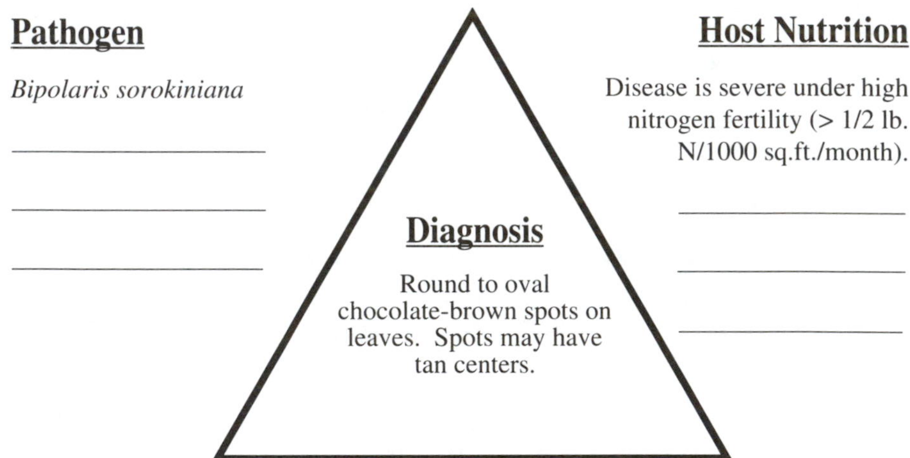

Pathogen

Bipolaris sorokiniana

Host Nutrition

Disease is severe under high nitrogen fertility (> 1/2 lb. N/1000 sq.ft./month).

Diagnosis

Round to oval chocolate-brown spots on leaves. Spots may have tan centers.

Environment

Temperatures of 77°-95°F (25°-35°C).
Disease severity increases with increase in temperature.
More than 10 hrs. of leaf wetness per day for several days.

Leaf Spot

Disease Forecasting and Pathogen Detection

A leaf spot forecaster and detection kits are not available.

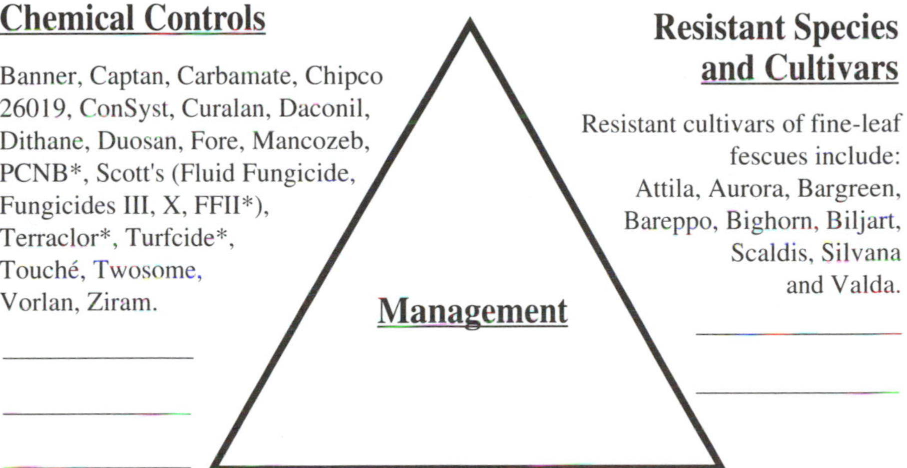

Chemical Controls

Banner, Captan, Carbamate, Chipco 26019, ConSyst, Curalan, Daconil, Dithane, Duosan, Fore, Mancozeb, PCNB*, Scott's (Fluid Fungicide, Fungicides III, X, FFII*), Terraclor*, Turfcide*, Touché, Twosome, Vorlan, Ziram.

Management

Resistant Species and Cultivars

Resistant cultivars of fine-leaf fescues include: Attila, Aurora, Bargreen, Bareppo, Bighorn, Biljart, Scaldis, Silvana and Valda.

Cultural Controls

Apply moderate amounts of nitrogen during summer (1/4-1/2 lb. N/ 1000 sq.ft./month).
Maintain moderate to high levels of soil P and K.
Decrease shade and increase air circulation to enhance drying of turf.
Avoid irrigation in late afternoon and in evening prior to midnight.
Limit thatch to 1/4 inch or less.
Raise mowing height.
Use light-weight mowing equipment to reduce stress.

*Avoid application to bentgrasses.

Necrotic Ring Spot

Hosts

Kentucky bluegrass and fine-leaf fescues.

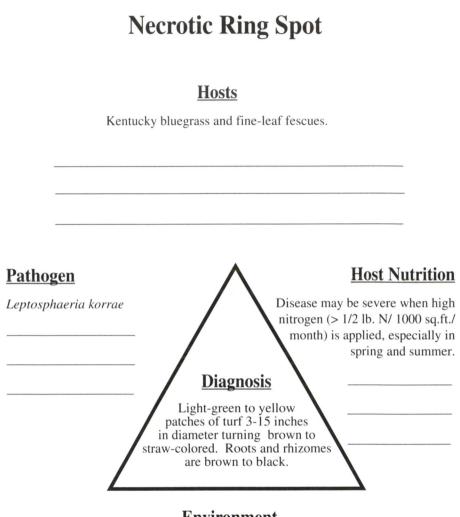

Pathogen

Leptosphaeria korrae

Host Nutrition

Disease may be severe when high nitrogen (> 1/2 lb. N/ 1000 sq.ft./ month) is applied, especially in spring and summer.

Diagnosis

Light-green to yellow patches of turf 3-15 inches in diameter turning brown to straw-colored. Roots and rhizomes are brown to black.

Environment

Disease initiated at temperature < 80˚F (26˚C) in moist soil.
Severity of symptoms increases with drought and high temperatures (> 80˚F).
Disease is severe on compacted soils.

Necrotic Ring Spot

Disease Forecastingand Pathogen Detection

A necrotic ring spot forecaster and detection kits are not available.

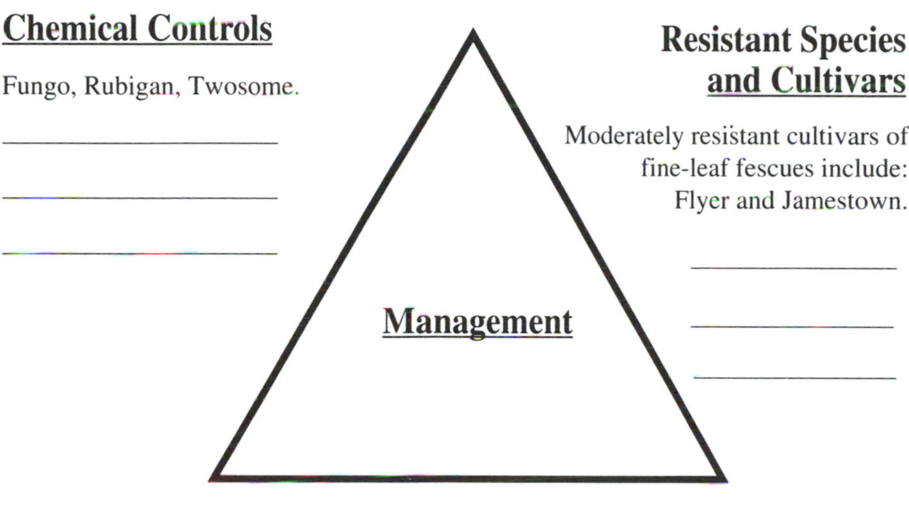

Chemical Controls

Fungo, Rubigan, Twosome.

Resistant Species and Cultivars

Moderately resistant cultivars of fine-leaf fescues include: Flyer and Jamestown.

Cultural Controls

Avoid high amounts of fast-release nitrogen.
Maintain moderate to high levels of phosphorus and potash according to soil tests.
Avoid drought stress.
Raise mowing height to at least 2 inches.
Avoid soil compaction.
Top-dress and aerify as needed, use light-weight equipment.
Reduce thatch thickness to 1/2 inch or less.

Nematodes

Hosts

All common species of turfgrasses.

Pathogen

More than 25 different species attack cool season turfgrasses.

Host Nutrition

Symptoms are often severe on lush, excessively fertilized turf or on thin, under-fertilized turf.

Diagnosis

Irregular shaped light green or yellow patches of turf up to several feet in diameter. Leaves may be yellow or brown from the tip. Roots may be thin, stunted or knotted.

Environment

Soil temperatures > 40˚F (5˚C).
Symptoms are often severe on turf growing in sandy, light-textured soils.
Symptoms may be enhanced by drought and high temperatures (>80˚F, 26˚C).

Nematodes

Disease Forecasting and Pathogen Detection

A nematode forecaster and detection kits are not available.

Chemical Controls

Post-plant nematicides:
Clandosan 618, Dasanit, Mocap*, Nemacur, Scott's Nematicide/ Insecticide.
Pre-plant nematicides:
Basamid, Brom-o-Sol, Telone, Terr-o-Cide, Terr-o-Gas, Vapam, Vorlex.

Resistant Species and Cultivars

Information on resistance among species of cool season turfgrasses is limited.

Management

Cultural Controls

Maintain a balanced fertility program.
Apply 1/2 lb. N/1000 sq.ft./month during spring and summer.
Maintain moderate to high levels of phosphorus and potash according to soil tests.
Have soil analyzed for nematodes prior to seeding or sodding.
Use sod that is nematode-free.

*Avoid applications to bentgrasses.

Net Blotch

Hosts

Fescues, bluegrasses and ryegrasses.

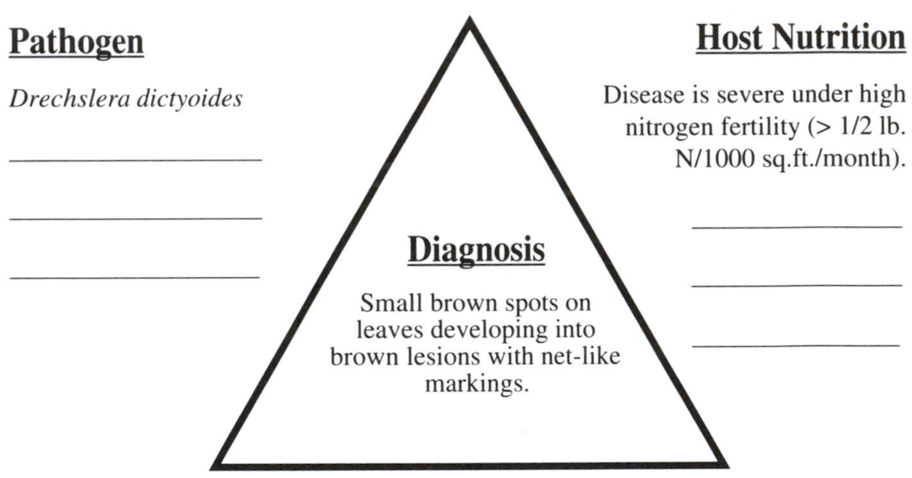

Pathogen

Drechslera dictyoides

Host Nutrition

Disease is severe under high nitrogen fertility (> 1/2 lb. N/1000 sq.ft./month).

Diagnosis

Small brown spots on leaves developing into brown lesions with net-like markings.

Environment

Cool temperatures of 40-80°F (5-27°C).
More than 10 hrs. of leaf wetness per day for several days.
Mowing height < 2 inches.

Net Blotch

Disease Forecasting and Pathogen Detection

A net blotch forecaster and detection kits are not available.

Chemical Controls

Banner, Chipco 26019, ConSyst, Curalan, Daconil, Fore, Mancozeb, Terraclor*, Turfcide*, Touché, Twosome, Vorlan.

Management

Resistant Species and Cultivars

Information on resistance among cultivars of fine-leaf fescues is limited.

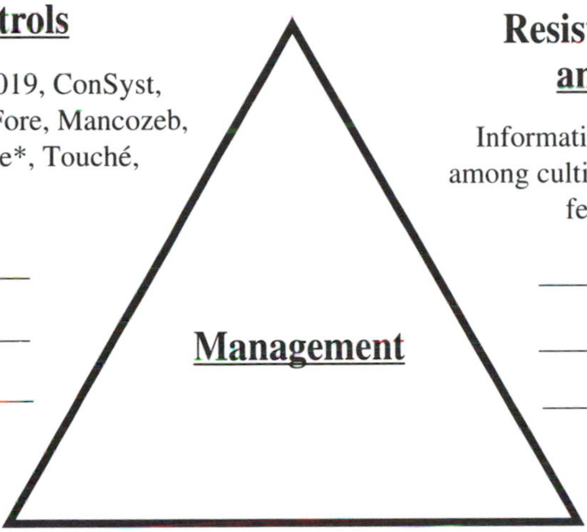

Cultural Controls

Fertilize with low to moderate levels of N during spring, summer and fall (1/4-1/2 lb. N/1000 sq.ft./month).
Decrease shade and increase air curculation to enhance drying of turf.
Avoid irrigation in late afternoon and in evening prior to midnight.
Raise mowing height to at least 2 inches.
Use light-weight mowing equipment to avoid stress on turf.

*Avoid applications to bentgrasses.

Nigrospora Blight

Hosts

Kentucky bluegrass, fescues and perennial ryegrass.

Pathogen

Nigrospora sphaerica

Host Nutrition

Disease may be severe on low nutrient turf.

Diagnosis

Straw-colored leaf lesions usually starting at tip forming large blighted areas of turf or circular patches 4-8 inches in diameter. White mycelium may form on leaves.

Environment

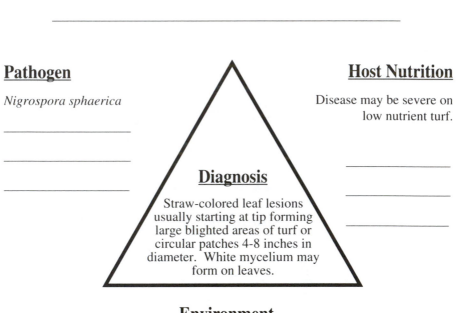

Temperature > 80°F (27°C).
More than 10 hrs. of leaf wetness per day for several days.
Disease may be severe on turf subjected to drought, herbicide stress or low mowing height.

Nigrospora Blight

Disease Forecasting and Pathogen Detection

A Nigrospora blight forecaster and detection kits are not available.

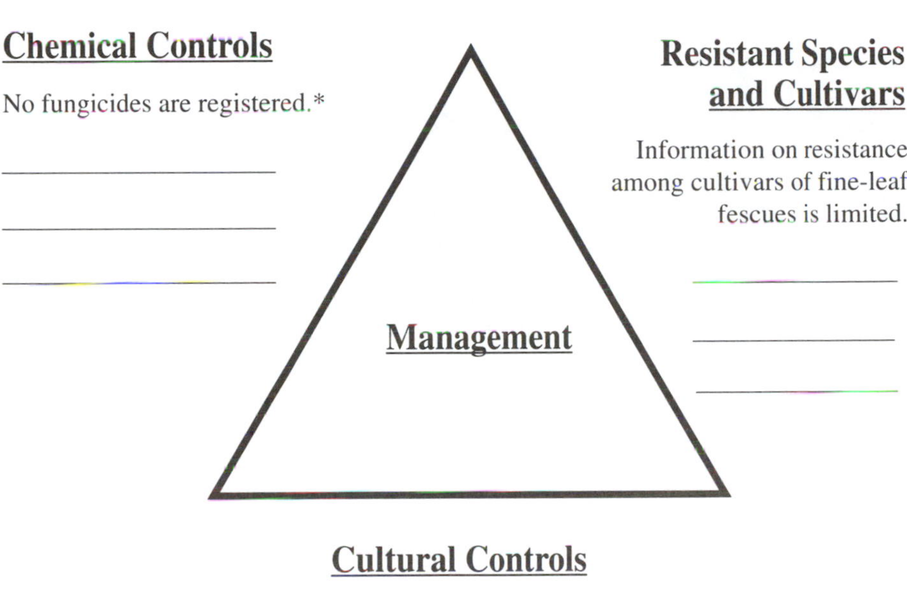

Chemical Controls

No fungicides are registered.*

Resistant Species and Cultivars

Information on resistance among cultivars of fine-leaf fescues is limited.

Management

Cultural Controls

Maintain moderate nitrogen fertility (1/2 lb. N/1000 sq.ft./month) during summer.
Maintain moderate to high levels of phosphorus and potash according to soil tests.
Avoid drought stress.
Avoid irrigation in late afternoon and evening prior to midnight.
Decrease shade and increase air circulation to enhance drying of turf.
Maintain turf at height of 2 inches or greater.
Avoid herbicide application during summer.

*Chipco 26019 and Daconil 2787 provide control in experimental evaluations.

Pink Patch

Hosts

Perennial ryegrass, fine-leaf fescues, bentgrasses and bluegrasses.

Pathogen

Limonomyces roseipellis

Host Nutrition

Disease is severe under conditions of low nitrogen fertility (< 1/2 lb. N/1000 sq.ft./month).

Diagnosis

Circular or irregular-shaped patches of tan-colored turf usually < 1 ft. in diameter. Red or pink mycelium grows from margin and tips of affected leaves.

Environment

Cool temperatures of 60°-75°F (16°-24°C).
More than 10 hrs. of leaf wetness per day for several days.

Pink Patch

Disease Forecasting and Pathogen Detection

A pink patch forecaster and detection kits are not available.

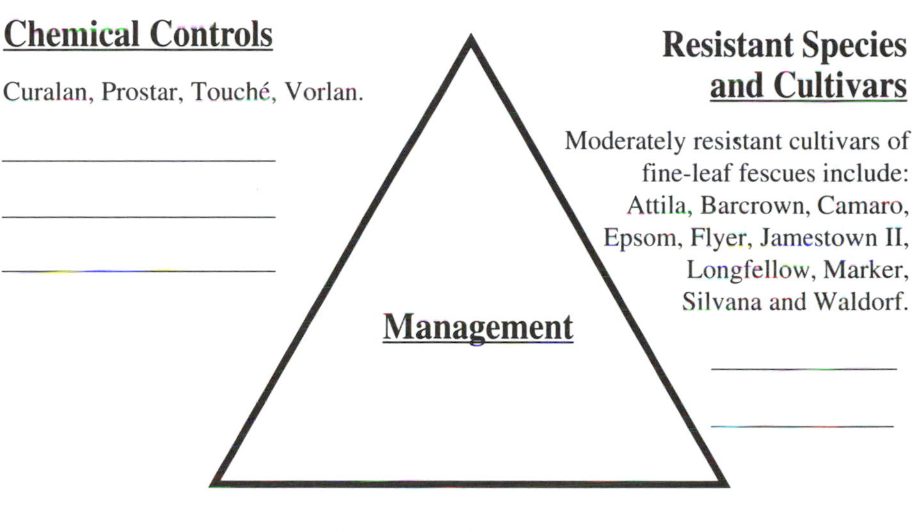

Chemical Controls

Curalan, Prostar, Touché, Vorlan.

Resistant Species and Cultivars

Moderately resistant cultivars of fine-leaf fescues include: Attila, Barcrown, Camaro, Epsom, Flyer, Jamestown II, Longfellow, Marker, Silvana and Waldorf.

Cultural Controls

Avoid low fertility.
Apply at least 1/2 lb. N/1000 sq.ft./month.
Maintain moderate to high levels of phosphorus and potash according to soil tests.
Reduce shade and increase air curculation to enhance drying of turf.
Avoid irrigation in late afternoon or in evening prior to midnight.
Mow turf at least once per week to remove diseased portions of leaf blades.

Pink Snow Mold

Hosts

All cool season grasses. Bentgrasses, annual bluegrass and perennial ryegrass are particularly susceptible.

Pathogen

Microdochium nivale (same pathogen causes Fusarium Patch)

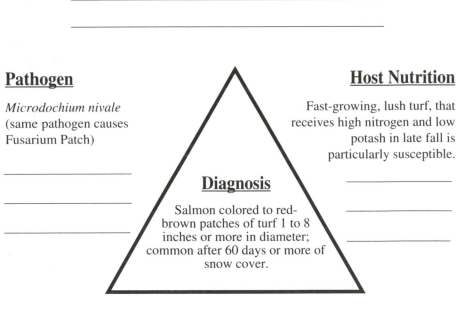

Host Nutrition

Fast-growing, lush turf, that receives high nitrogen and low potash in late fall is particularly susceptible.

Diagnosis

Salmon colored to red-brown patches of turf 1 to 8 inches or more in diameter; common after 60 days or more of snow cover.

Environment

Disease is common after at least 60 days of snow cover, but pathogen can infect turf in absence of snow (see Fusarium Patch).
Disease is particularly severe when snow covers unfrozen ground.

Pink Snow Mold

Disease Forecasting and Pathogen Detection

A pink snow mold forecaster and detection kits are not available.

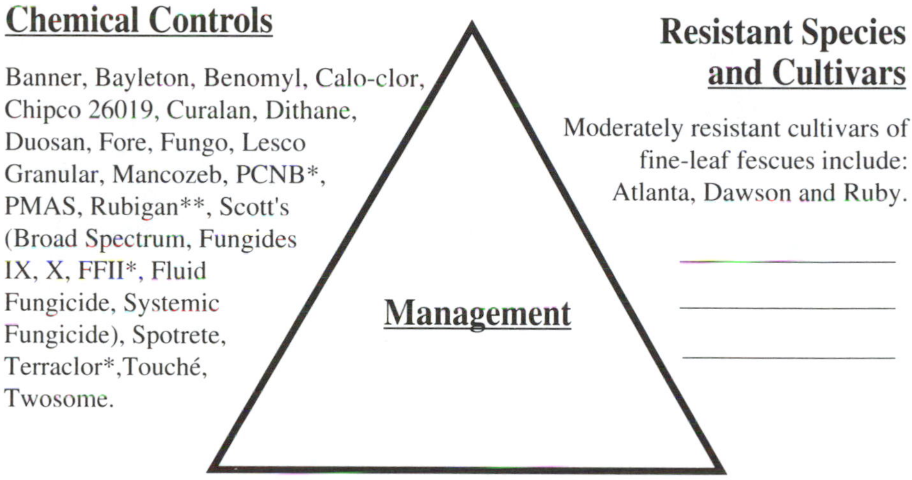

Chemical Controls

Banner, Bayleton, Benomyl, Calo-clor, Chipco 26019, Curalan, Dithane, Duosan, Fore, Fungo, Lesco Granular, Mancozeb, PCNB*, PMAS, Rubigan**, Scott's (Broad Spectrum, Fungides IX, X, FFII*, Fluid Fungicide, Systemic Fungicide), Spotrete, Terraclor*,Touché, Twosome.

Resistant Species and Cultivars

Moderately resistant cultivars of fine-leaf fescues include: Atlanta, Dawson and Ruby.

Cultural Controls

Maintain moderate nitrogen fertility (1/2 lb. N/1000 sq.ft./month) during late summer and fall.
Maintain high potash levels according to soil tests.
Use snow fence, shrubs or knolls as windbreaks to prevent excess snow from accumulating.
Prevent snow compaction by machinery or skiers.
Melt snow in spring with organic fertilizers.
Physically remove snow in spring.
Follow controls for Fusarium Patch after snow melt.

*Avoid application to actively growing bentgrasses. **May reduce populations of annual bluegrass.

Powdery Mildew

Hosts

Kentucky bluegrass and fine-leaf fescues.

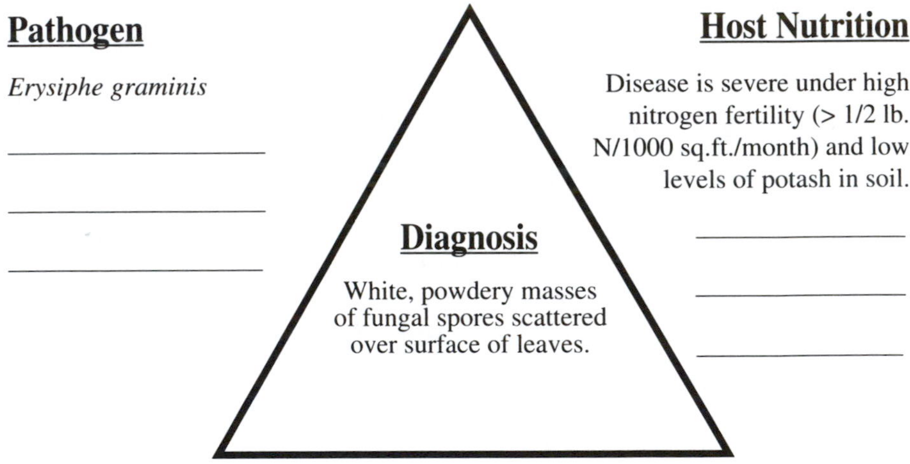

Pathogen

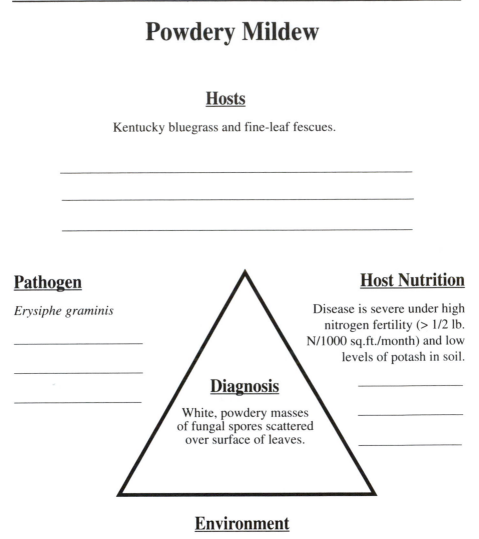

Erysiphe graminis

Host Nutrition

Disease is severe under high nitrogen fertility (> 1/2 lb. N/1000 sq.ft./month) and low levels of potash in soil.

Diagnosis

White, powdery masses of fungal spores scattered over surface of leaves.

Environment

Disease is severe in shaded areas at temperatures of 60°-72°F (15°-22°C). High humidity is required for infection, but leaf wetness is not essential.

Powdery Mildew

Disease Forecasting and Pathogen Detection

A powdery mildew forecaster and detection kits are not available.

Chemical Controls

Banner, Bayleton, Lesco Granular and Systemic, Rubigan, Twosome.

Resistant Species and Cultivars

Information on resistance among cultivars of fine-leaf fescues is limited.

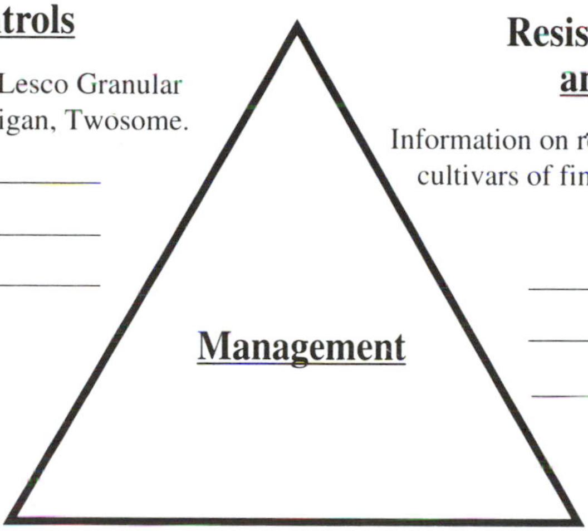

Management

Cultural Controls

Maintain moderate nitrogen fertility (1/2 lb. N/1000 sq.ft./month) and moderate to high levels of potash according to soil tests.
Reduce shade and increase air circulation.

Pythium Blight

Hosts

All cool season grasses. Annual bluegrass and perennial ryegrass are particularly susceptible.

Pathogen

Pythium aphanidermatum and other species of *Pythium.*

Host Nutrition

Disease is severe under high nitrogen fertility (> 1/2 lb. N/1000 sq.ft./month). Deficiency in calcium may increase susceptibility.

Diagnosis

Greasy brown patches of turf an inch or less in diameter, increasing to several inches and turning straw colored. Grey-white, cottony mycelium observed in early morning.

Environment

Night temperature > 65°F (18°C).
More than 10 hrs. of leaf wetness per day for several days.
Poor surface and sub-surface drainage.

Pythium Blight

Disease Forecasting and Pathogen Detection

Pythium blight forecasters are available from Pest Management Supply, P.O. Box 936, Amherst, MA 01004 or Neogen Corp., 620 Lesher Pl., Lansing, MI 48912. Detection kits are available from Neogen Corp.

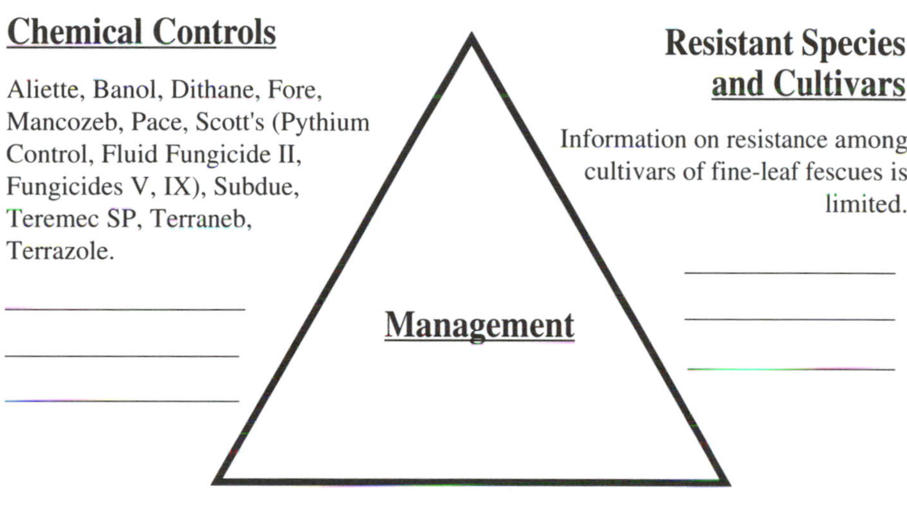

Chemical Controls

Aliette, Banol, Dithane, Fore, Mancozeb, Pace, Scott's (Pythium Control, Fluid Fungicide II, Fungicides V, IX), Subdue, Teremec SP, Terraneb, Terrazole.

Management

Resistant Species and Cultivars

Information on resistance among cultivars of fine-leaf fescues is limited.

Cultural Controls

Maintain moderate nitrogen fertility (1/2 lb. N/1000 sq.ft./month).
Maintain optimum plant calcium levels.
Decrease shade and increase air circulation to enhance drying of turf.
Improve surface and subsurface drainage.
Avoid mowing susceptible areas when turf is wet, particularly when night temperatures are > 70°F (21°C).

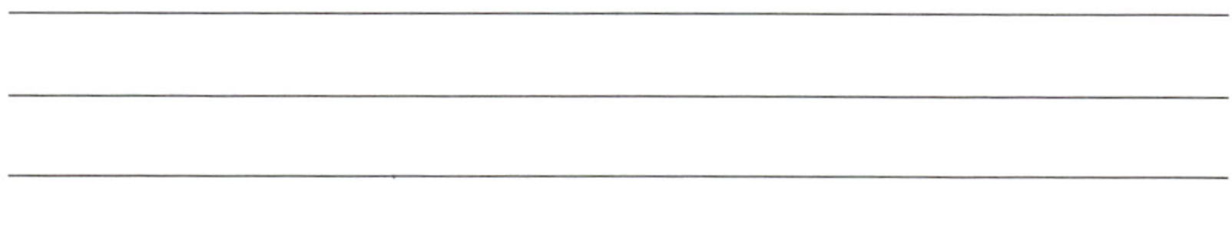

Pythium Root Rot

Hosts

All species of cool season grasses.

Pathogen

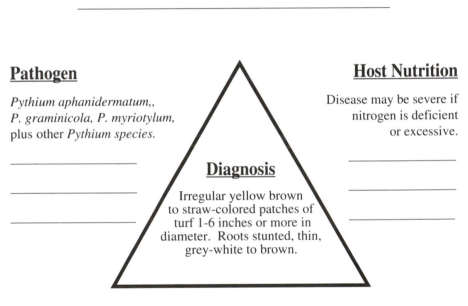

Pythium aphanidermatum,,
P. graminicola, P. myriotylum,
plus other *Pythium species.*

Host Nutrition

Disease may be severe if nitrogen is deficient or excessive.

Diagnosis

Irregular yellow brown to straw-colored patches of turf 1-6 inches or more in diameter. Roots stunted, thin, grey-white to brown.

Environment

Cool (32°-50°F, 0-10°C) or warm (70°-90°F, 21-32°C) soil temperatures*.
High soil moisture.
Poor surface or subsurface drainage.
Conditions unfavorable for carbohydrate development by leaves – low light, low mowing height, excessive wear.

* Some *Pythium* species are favored by cool soils, other species by warm soils.

Pythium Root Rot

Disease Forecasting and Pathogen Detection

A Pythium root rot forecaster is not available. Detection kits are available from Neogen Corp., 620 Lesher Pl., Lansing, MI 48912.

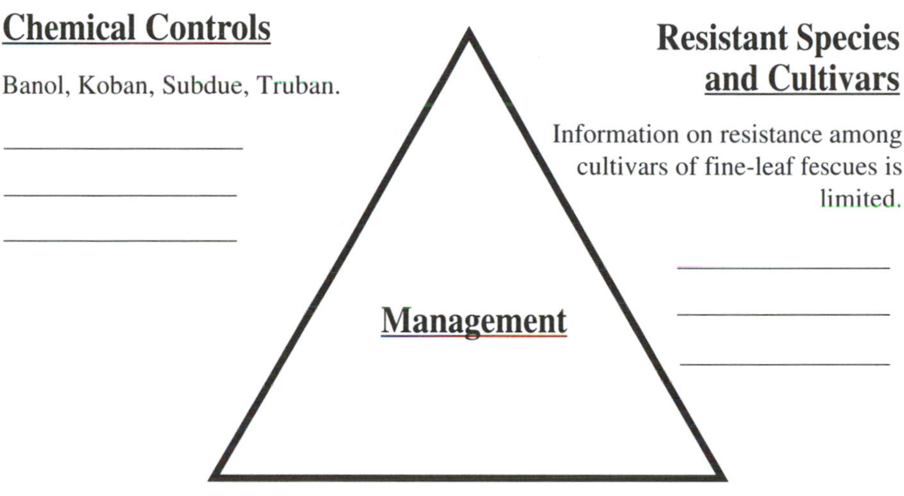

Chemical Controls

Banol, Koban, Subdue, Truban.

Resistant Species and Cultivars

Information on resistance among cultivars of fine-leaf fescues is limited.

Cultural Controls

Maintain moderate levels of nitrogen (1/2 lb. N/1000 sq.ft./month).
Do not over fertilize with nitrogen in spring when roots are forming.
Maintain moderate to high levels of phosphorus and potash according to soil tests.
Improve surface and subsurface drainage.
Raise mowing height.
Decrease shade.
Use light-weight mowing equipment.
Applications of certain composts may reduce disease severity.

Red Thread

Hosts

Bentgrasses, bluegrasses, fine-leaf fescues and perennial ryegrass. Fine leaf fescues and perennial ryegrass are particularly susceptible.

Pathogen

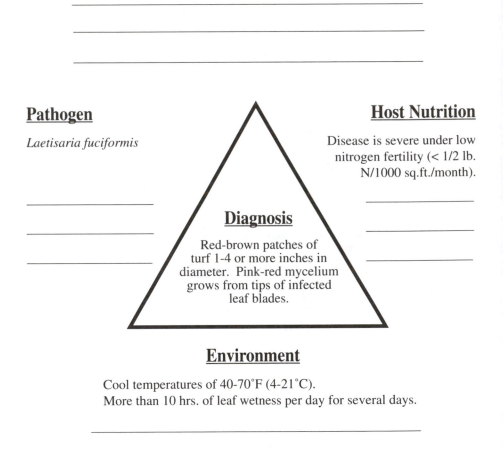

Laetisaria fuciformis

Host Nutrition

Disease is severe under low nitrogen fertility (< 1/2 lb. N/1000 sq.ft./month).

Diagnosis

Red-brown patches of turf 1-4 or more inches in diameter. Pink-red mycelium grows from tips of infected leaf blades.

Environment

Cool temperatures of 40-70˚F (4-21˚C).
More than 10 hrs. of leaf wetness per day for several days.

Red Thread

Disease Forecasting and Pathogen Detection

A red thread forecaster and detection kits are not available.

Chemical Controls

Banner, Bayleton, Chipco 26019, Cleary's 3336, ConSyst, Curalan, Daconil, Dithane, Duosan, Fore, Fungo, Lesco Granular and Systemic, Mancozeb, Prostar, Rubigan*, Touché, Twosome, Vorlan.

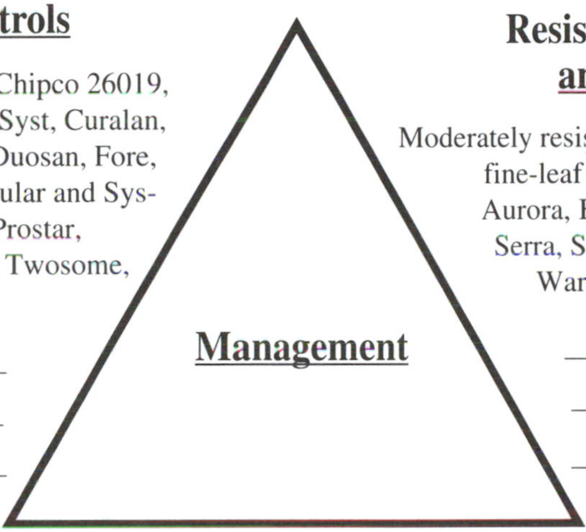

Resistant Species and Cultivars

Moderately resistant cultivars of fine-leaf fescues include: Aurora, Brigade, Reliant, Serra, Silvana, Waldorf, Warwick and Valda.

Management

Cultural Controls

Fertilize with at least 1/2 lb. N/1000 sq.ft./month.
Maintain moderate to high levels of potash and phosphorus according to soil tests.
Reduce shade and increase air circulation to enhance drying of turf.
Avoid irrigation in late afternoon or in evening prior to midnight.
Maintain soil pH at 6.5 to 7.0.
Mow turf at least once per week to remove diseased portions of leaf blades.

*May reduce populations of annual bluegrass.

Rusts

Hosts

All common species of cool season grasses.

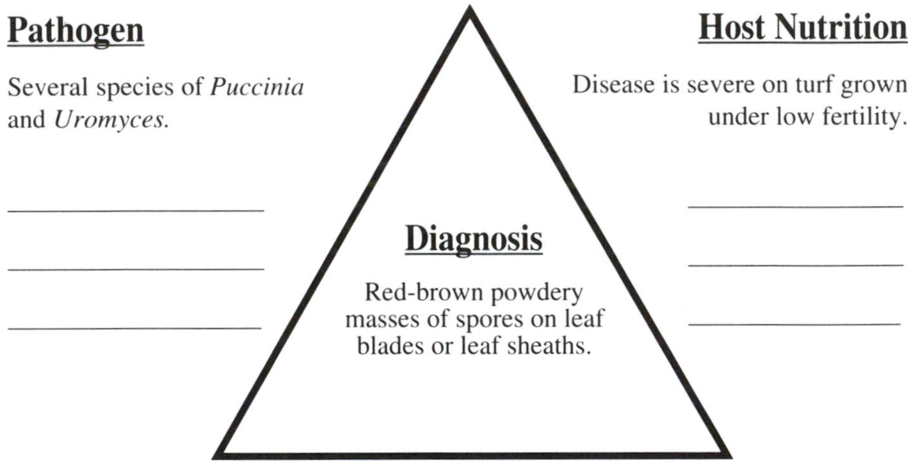
Pathogen
Several species of *Puccinia* and *Uromyces*.
Host Nutrition
Disease is severe on turf grown under low fertility.
Diagnosis
Red-brown powdery masses of spores on leaf blades or leaf sheaths.

Environment

Temperatures of 68°-86°F (20°-30°C).
Disease is severe on turf subjected to drought stress, low mowing, shade or poor air circulation.

Rust

Disease Forecasting and Pathogen Detection

A rust forecaster and detection kits are not available.

Chemical Controls

Banner, Bayleton, Captan, Carbamate, Cleary's 3336, ConSyst, Daconil, Dithane, Duosan, Fore, Mancozeb, Rubigan**, Scott's (Fluid Fungicide III, Fungicide VII, FFII*), Twosome, Ziram.

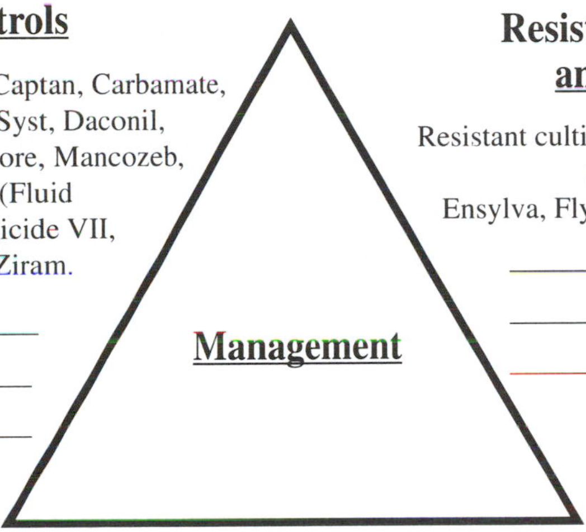

Resistant Species and Cultivars

Resistant cultivars of fine-leaf fescues include: Ensylva, Flyer and Shadow.

Management

Cultural Controls

Maintain moderate and balanced fertility throughout growing season.
Reduce shade and increase air circulation to enhance drying of turf.
Increase mowing height.
Avoid drought stress.
Avoid irrigation in late afternoon and in early evening prior to midnight.

*Avoid application to bentgrasses. **May reduce populations of annual bluegrass.

Summer Patch

Hosts

Bluegrasses and fine-leaf fescues.

Pathogen

Magnaporthe poae

Host Nutrition

Disease may be severe when turf is fertilized with fast-release sources of nitrogen.

Diagnosis

Circular patches of wilted to straw-colored turf, usually less than 10 inches in diameter. Leaves turn yellow or brown starting at tips. Roots are light to dark brown.

Environment

Day-time temperature > 85°F (29°C).
High soil moisture.
Poor surface or subsurface drainage.
Low mowing height.

Summer Patch

Disease Forecasting and Pathogen Detection

A summer patch forecaster and detection kits are not available.

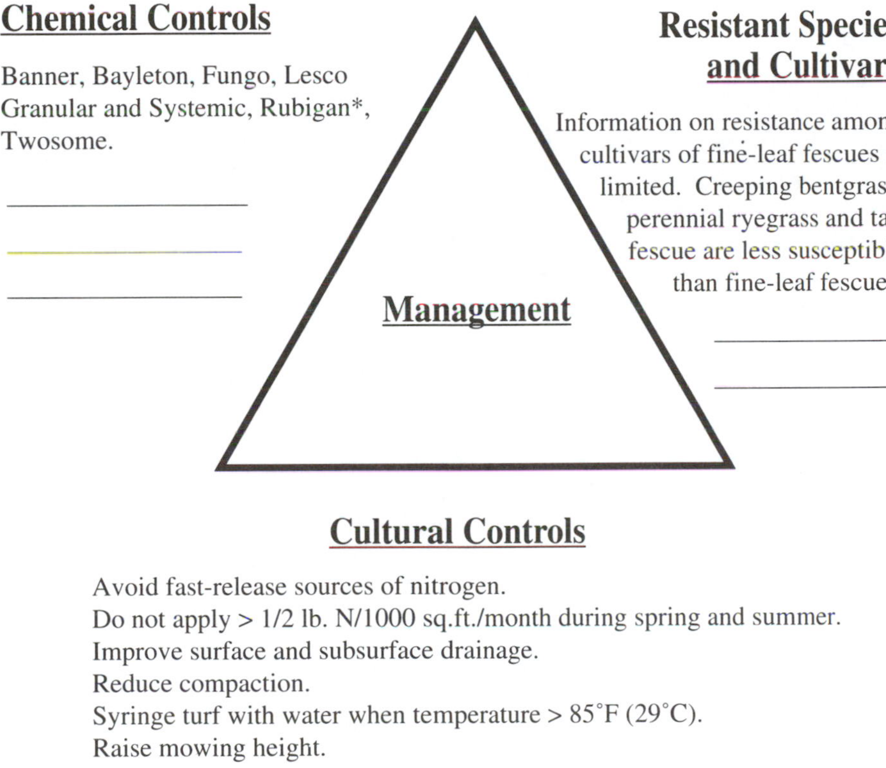

Chemical Controls

Banner, Bayleton, Fungo, Lesco Granular and Systemic, Rubigan*, Twosome.

Resistant Species and Cultivars

Information on resistance among cultivars of fine-leaf fescues is limited. Creeping bentgrass, perennial ryegrass and tall fescue are less susceptible than fine-leaf fescues.

Cultural Controls

Avoid fast-release sources of nitrogen.
Do not apply > 1/2 lb. N/1000 sq.ft./month during spring and summer.
Improve surface and subsurface drainage.
Reduce compaction.
Syringe turf with water when temperature > 85˚F (29˚C).
Raise mowing height.
Use light-weight mowing equipment.

*May reduce populations of annual bluegrass.

Typhula Blight
(Gray Snow Mold)

Hosts

All cool-season turfgrasses. Bentgrasses, annual bluegrass and perennial ryegrass are particularly susceptible.

Pathogen

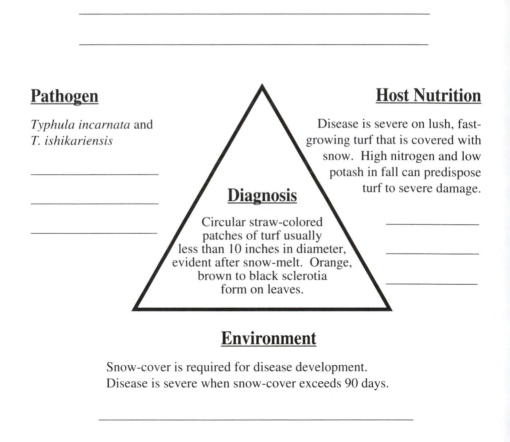

Typhula incarnata and *T. ishikariensis*

Host Nutrition

Disease is severe on lush, fast-growing turf that is covered with snow. High nitrogen and low potash in fall can predispose turf to severe damage.

Diagnosis

Circular straw-colored patches of turf usually less than 10 inches in diameter, evident after snow-melt. Orange, brown to black sclerotia form on leaves.

Environment

Snow-cover is required for disease development. Disease is severe when snow-cover exceeds 90 days.

Typhula Blight

Disease Forecasting and Pathogen Detection

A Typhula blight forecaster and detection kits are not available.

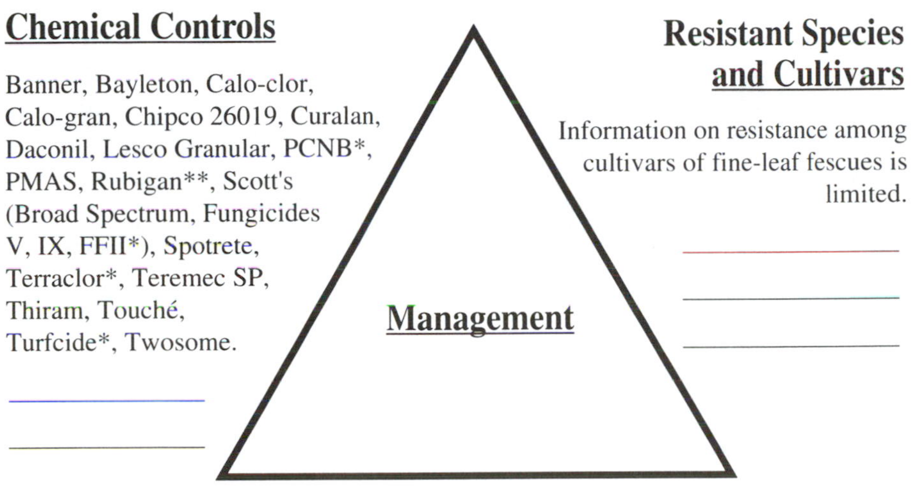
Chemical Controls
Banner, Bayleton, Calo-clor, Calo-gran, Chipco 26019, Curalan, Daconil, Lesco Granular, PCNB*, PMAS, Rubigan**, Scott's (Broad Spectrum, Fungicides V, IX, FFII*), Spotrete, Terraclor*, Teremec SP, Thiram, Touché, Turfcide*, Twosome.
Management
Resistant Species and Cultivars
Information on resistance among cultivars of fine-leaf fescues is limited.

Cultural Controls

Avoid a fertility program that results in lush, fast-growing turf in late fall and winter.
Maintain high potash levels according to soil tests.
Use snow fence, hedges or knolls to prevent snow from accumulating excessively on turf.
Use dark-colored organic fertilizers or composts to melt snow in spring.
Physically remove snow in spring.
Prevent compaction of snow during winter.

*Avoid application to actively growing bentgrasses. **May reduce populations of annual bluegrass.

White Patch

(White Blight)

Hosts

Fescues, bluegrasses and creeping bentgrass.

Pathogen

Melanotus phillipsi

Host Nutrition

Effects of nutrition on disease are unknown.

Diagnosis

Patches of bleached white turf a few inches to over one foot in diameter. Leaves are bleached white from tip down. Mushroom fruiting bodies may be observed on leaves.

Environment

Night temperatures > 70˚F (21˚C).
More than 10 hrs. of leaf wetness per day for several days.
Disease is particularly severe on soils from recently cleared forests.

White Patch

Disease Forecasting and Pathogen Detection

A white-patch forecaster and detection kits are not available.

Chemical Controls

No fungicides are registered. Fungicides registered for stripe smut, Typhula blight or brown patch may suppress disease.

Resistant Species and Cultivars

Information on resistance among cultivars of fine-leaf fescues is limited.

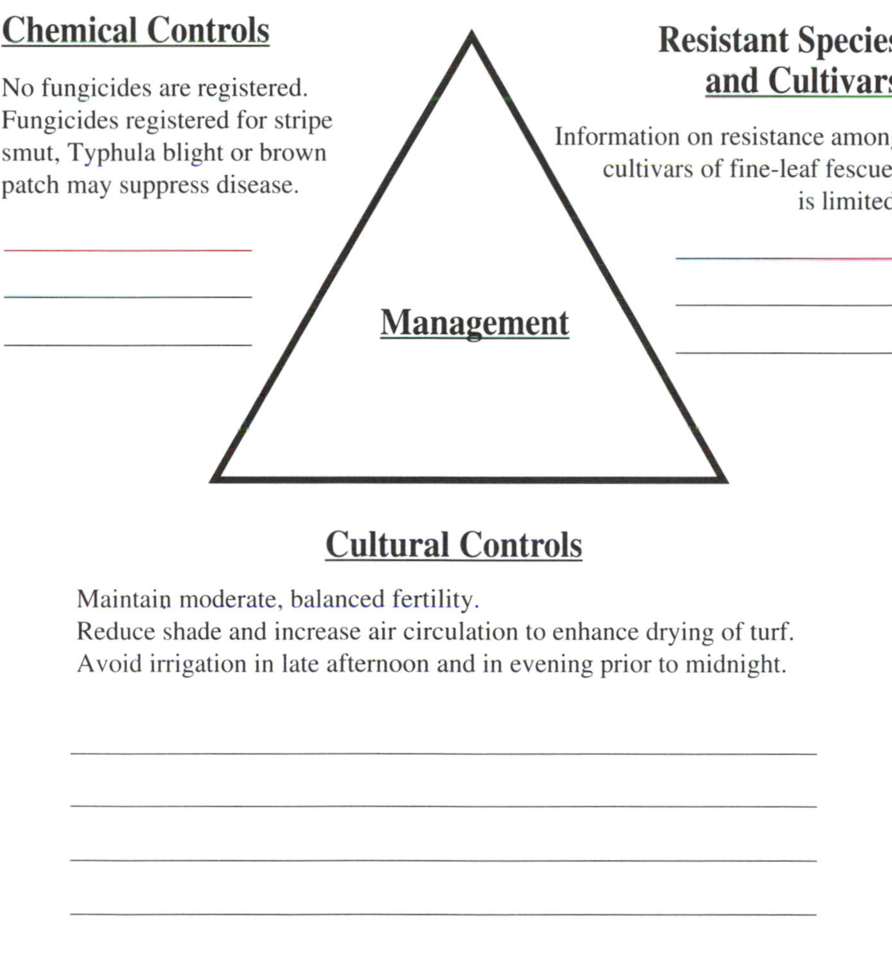

Cultural Controls

Maintain moderate, balanced fertility.
Reduce shade and increase air circulation to enhance drying of turf.
Avoid irrigation in late afternoon and in evening prior to midnight.

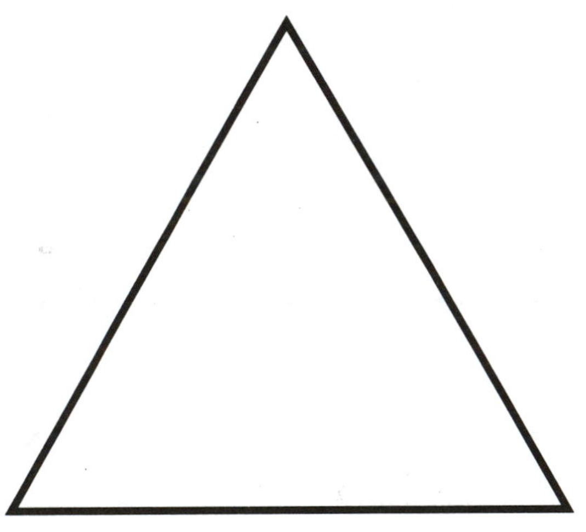